COLLAPSE OF TORI AND GENESIS OF CHAOS IN DISSIPATIVE SYSTEMS

KUNIHIKO KANEKO

World Scientific

Published by

World Scientific Publishing Co Pte Ltd.
P. O. Box 128, Farrer Road, Singapore 9128

Library of Congress Cataloging-in-Publication Data is available.

ISBN 9971-978-61-X

Printed in Singapore by Kyodo-Shing Loong Printing Industries Pte Ltd.

FOREWORD

Since ancient times, people have recognized the importance of chaos, while they have searched for the order. They have thought that the chaos may be the source for the creativity and life, as can be seen in a lot of the great literature or the masterpieces of philosophers.

Now, the nonlinear science, above all, chaos is vivid and revealing novel aspects in nature every day. The purpose of the present book however, is not to tell the established knowledge on chaos but to let you show some aspects in the frontier of chaos research and to tell what have not yet been understood and what may lie in the future. If the present book can be a clue to your new step in nonlinear science, it is of my greatest pleasure.

The present book is based on my thesis "Collapse of Tori and Genesis of Chaos in Nonlinear Nonequilibrium Systems" submitted to the University of Tokyo in December 1983 for the requirement of Ph.D. degree, though some parts are newly written here.

For the completion of the thesis, I would like to thank the following people for fruitful discussions, encouragements and invaluable comments: Prof. M. Suzuki (thesis advisor at Univ. of Tokyo), Prof. K. Tomita, Prof. Y. Kuramoto, Dr. Y. Aizawa, Dr. K. Ikeda, Dr. I. Tsuda, Dr. H. Daido (Kyoto Univ. group), Prof. H. Mori (Kyushu Univ.), Dr. M. Sano, Prof. Y. Sawada (Tohoku Univ.), Dr. S. Takesue and the members of vi

research group at University of Tokyo, Prof. A. Libchaber (Chicago), Prof. D. Rand (Warwick), Prof. H.L. Swinney (Texas), Prof. Y. Takahashi, Prof. T. Izuyama, and Prof. M. Wadati (Institute of Physics, Univ. of Tokyo).

The following parts are largely enlarged from the thesis; §2.7, 2.8*, 2.9, 5.2, 6.4*, 8.1, 8.2, 8.3 (* ... new section). Chapter 7 is new, which is an enlarged version of §7.3 in my thesis. For other sections, only the recent progress (up to June, 1985) is amended if necessary. For this enlargement, I am also grateful to Drs. J.D. Farmer, G. Mayer-Kress, J. Keeler, R. Deissler, D. Umberger, R. Bagley (Los Alamos), J. Crutchfield (Berkeley), N. Packard, and S. Wolfram (Princeton), for stimulating discussions, critical comments, and hospitality during my stay at Los Alamos. I would also like to thank Prof. C. Jeffries (Berkeley), Prof. A. Arneodo (Nice), and Prof. B.A. Huberman (Xerox) for sending me the preprints prior to publications.

For Chapters 1, 7, and 8, I have largely benefitted from the discussions with Dr. I. Tsuda (Bioholonics), Dr. I. Shimada (Nihon Univ.), Mr. Y. Iba, (Univ. of Tokyo), Dr. Y. Aizawa, and Dr. K. Ikeda (Kyoto Univ.). I would like to express my sincere thanks to these people. For the publication, I am largely indebted to Dr. K.K. Phua (World Scientific).

Hoping that the present book can help to excite the chaos in your brain and keep it active.

金子　邦彦
KANEKO Kunihiko

1985 July

ACKNOWLEDGEMENTS

The publisher would like to thank the following authors and publishers for permission to reproduce or modify copyright materials (figure numbers within brackets refer to this publication). Credit goes to the following:

E.N. Lorenz, J. Atmos. Sci. 20 (1983) 130, American Meteorological Society (Figs. 1.1-1.3); B.B. Mandelbrot, The Fractal Geometry of Nature, W.H. Freeman and Co. (Fig. 1.4); M. Hénon, Comm. Math. Phys. 50 (1976) 69, Springer-Verlag (Fig. 1.5); A. Libchaber, C. Laroche and S. Fauve, J. de Phys. L43 (1982) (Fig. 1.6); K. Kaneko, "Mechanism of Frequency Locking ..." in Turbulence and Chaotic Phenomena in Fluids (1985), North-Holland (Fig. 2.7); S. Ostlund, D. Rand, J. Sethna and E.D. Siggia, Physica 8D (1983) 303, North-Holland (Figs. 2.14-2.15); M.H. Jensen, P. Bak and T. Bohr, Phys. Rev. Lett. 50 (1983) 1637 (Fig. 2.16); J.P. Gollub and S.V. Benson, J. Fluid Mech. 100 (1980) 449 (Figs. 2.27-2.28 & 6.11); R.V. Buskirk and C. Jeffries, Phys. Rev. A31 (1985) 3332 (Figs. 2.29-2.31 & 4.13); P. Bergé, Physica Scripta T1 (1982) 71 (Fig. 4.1); M. Sano and Y. Sawada, Chaos and Statistical Methods ed. Y. Kuramoto (1983), Springer-Verlag (Fig. 5.10); A. Libchaber, S. Fauve and C. Laroche, Physica 7D (1983) 73, North-Holland (Fig. 6.12); C. Grebogi, E. Ott and J.A. Yorke, Physica 15D (1985) 354, North-Holland (Table 6-II); K. Kaneko, Dynamical Problems in Soliton Systems (1985) 272, Springer-Verlag (Figs. 7.1-7.5 & 7.7); K. Kaneko, Prog. Theor. Phys. (Figs. 2.4-2.6, 2.8-2.13, 2.18-2.25, 2.32-2.34, 3.1-3.11, 4.2-4.12, 5.1-5.9, 6.1-6.8, 7.8, Tables 5-I & 6-I).

© M.C. Escher Heirs c/o Cordon Art - Baarn - Holland for the Chapter opening figures 1-7.

To those who have not granted us permission before publication, we have taken the liberty to reproduce their figures without consent. We will however acknowledge them in future editions of this work.

CONTENTS

Chapter 1

INTRODUCTION

Order and Chaos
by *M. C. Escher*

§1.1 Dawn of Nonlinear Nonequilibrium Physics

Physics has built new paradigms[1] and has made progresses by enlarging its field of vision. The development of new technologies, such as black body radiation, opened the paradigm of quantum mechanics while the new world of electromagnetic waves opened the theory of relativity. In the field of statistical physics we have experienced various examples of new paradigms, such as linear response theory, superconductivity, and renormalization group theory of critical phenomena.

What is happening in our age? Has the construction of great new paradigms come to end? I don't think so. It seems to me, on the contrary, that we are now in the era of a large paradigm shift.[2]

The most remarkable feature of our age is the rapid developments of computers.[3] As is usually seen in history of sciences, quantitative changes induce qualitative changes, which cause a new paradigm shift. Then, what does the development of computers bring about? It seems to me that the paradigm of nonlinear physics has been brought about from numerical studies using computers.

The new world of nonlinear physics seemed to have been noticed by great physicists before many years ago. Poincaré had already known a great deal of important notions in nonlinear physics. M. Born pointed out the unpredictability in classical mechanics[4], in response to Einstein's criticism on the abandonment of the determinism in quantum mechanics. In 1963, W. Heisenberg wrote a review paper on nonlinear physics[5], where he insisted upon the importance of linearization, unpredictability, symmetry, statistical methods in nonlinear physics, which seems to anticipate the developments of today. In spite of their great perspectives, however, they could not proceed further without computers.

The new paradigm of nonlinear (nonequilibrium) physics is also required by other fields of sciences. Tomonaga pointed out in his last book that the "empire of physics" might split into smaller states, such as biophysics, geophysics, astrophysics etc. just as the Roman Empire did.[6] One of the most interesting and important common features of these fields is that systems in these fields are placed in nonlinear nonequilibrium states.[7] In these states, there appear new and remarkable structures, which were called "dissipative structures" by I. Prigogine.[8]

Examples of these structures are as follows;

1) pattern; steady structures which vary in space. Examples are convection of fluids, chemical reactions etc.;

2) rhythm (limit cycle); homogeneous states which vary periodically in time. Examples are chemical oscillations, circadian rhythm, heartbeat, etc.;

3) rhythm (quasiperiodic states); similar to 2), but the state varies quasiperiodically in time;

4) chaos; the state which varies chaotically in time;

5) turbulence; the difference between chaos and turbulence is not so clear, but the "turbulence" usually includes much more modes than chaos and the state changes chaotically both in time and in space;

etc. Rayleigh-Bénard convection and Taylor-Couette vortices show beautiful examples of the states 1)-5).

Prigogine's group has energetically studied the dissipative structures, mainly for the cases 1) and 2).[8] These works may give a basis on our understanding of biological systems. (Living states must be placed far from equilibrium, otherwise death visits according to the second law of thermodynamics[9]). These phenomenological studies are remarkably different from orthodox physics, in the point that a decomposition into elementary parts has never been looked for.[10] If we look back again on the history of the Roman Empire, important inheritance from an empire may always be the most heretical thing in the empire, just as Christianity in the Roman Empire.

Haken called the new paradigm as "synergetics"[11] and organized a series of conferences, in which a large number of scientists from various fields have participated and contributed to the formation of new paradigm.[12]

4

In recent years, a new type of structure, i.e., chaos (or turbulence) has attracted a lot of scientists, which will be described in later sections. The great progress in this field in recent years would have been impossible without computers. We can see a lot of examples for the important roles of computers in the history of chaos physics. In this sense, nonlinear physics, above all, chaos physics may symbolize our age of computers.

Though the recent greatest paradigm, quantum mechanics, is linear, it involves important notions in nonlinear physics. First, unpredictability was introduced by Heisenberg's uncertainty principle. The unpredictability in a macroscopic level was first introduced by chaos physics, where sensitive dependence on initial conditions play a very important role. Thus, the Laplace's determinism, denied in a microscopic level, has also been demolished in a macroscopic level, by the recent progress in nonlinear physics.

As for the revolt against the reductionism, the standpoint of nonlinear physics is rather similar to that of thermodynamics. However, we can see a germ of this revolt also in quantum mechanics. Since the observation in one place must always influence the observations in other places, complete reductionism is impossible. Revolts against reductionism in our age, in this sense, might stem from the observation theory of quantum mechanics.

In the following two sections, discovery of chaos in dissipative systems, dawn of chaos research, and the recent progress in the study on the onset of chaos will be briefly reviewed. Since they are rather introductory, those who know the "commonsense" of chaos physics can omit these two sections.

§1.2 Dawn of Chaos Physics

In the present book, we restrict ourselves only to the dissipative systems. Chaos in a dissipative system was first discovered by E.N. Lorenz.[13] Lorenz studied the problem of the thermal convection (his purpose was to analyze the motion of the atmosphere). He expanded a set of nonlinear partial differential equations (Navier-Stokes equation and the thermal equation) by Fourier transformations and truncated, to choose only three modes. The set of equations thus obtained, is the famous Lorenz model

$$\dot{X} = - \sigma(X - Y)$$

$$\dot{Y} = - XZ + rX - Y$$

$$\dot{Z} = XY - bZ \quad . \tag{1.2.1}$$

He found that the numerical solution of Eq. (1.2.1) for some parameter values (e.g., σ = 10, b = 8/3, r = 28) shows an irregular behavior (see Figs. 1.1 and 1.2). The solution has a sensitive dependence on initial conditions, i.e., a small difference of initial values of two orbits grow exponentially. He concluded, from this observation, that very-long-range weather forecasting is impossible.

His paper includes abundant new ideas and new methods. In Fig. 1.3, the abscissa is M_n, the value of the n-th maximum of Z, while the ordinate is M_{n+1}, the value of the following maximum. As can be seen from this figure, a one-dimensional map $M_n \rightarrow M_{n+1}$ is extracted from the original differential equations (1.2.1). This method is called Lorenz plot. The one-dimensional map in Fig. 1.3 is quite similar to the map

$$M_{n+1} = 2M_n \qquad (M_n \leq 1/2)$$

$$M_{n+1} = 2 - 2M_n \qquad (M_n \geq 1/2) \quad . \tag{1.2.2}$$

Sensitive dependence on initial conditions can be easily seen from the map (1.2.2). The method of Lorenz plot has been widely used in the study of chaos.

The importance of one-dimensional mappings was reviewed by R. May,[14] in connection with the population biology. In many problems of population biology (e.g., genetics, epidemology, ecology, and also social sciences), the change of a population is expressed by difference equations. A typical example is a logistic map

6

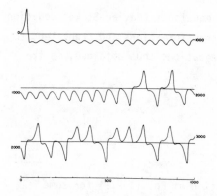

Fig. 1.1 Numerical solution of the convection equations. Graph of Y as a function of time for the first 1000 iterations (upper curve), second 1000 iterations (middle curve), and third 1000 iterations (lower curve).

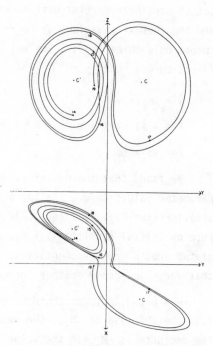

Fig. 1.2 Numerical solution of the convection equations. Projections on the X-Y-plane and the Y-Z-plane in phase space of the segment of the trajectory extending from iteration 1400 to iteration 1900. Numerals "14," "15," etc., denote positions at iterations 1400, 1500, etc. States of steady convection are denoted by C and C'.

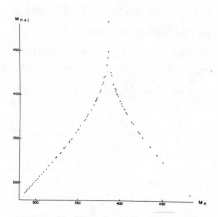

Fig. 1.3 Corresponding values of relative maximum of Z (abscissa) and subsequent relative maximum of Z (ordinate) occurring during the first 6000 iterations.

Figures 1.1-1.3 are cited from Lorenz's paper (Ref. 13).

$$x_{n+1} = ax_n(1 - x_n) \quad . \tag{1.2.3}$$

Though the continuous time approximation for the map (1.2.3) shows only a monotonic approach to a constant value, the original map (1.2.3) has very abundant phenomena. Various types of cycles (including unstable ones) appear according to the Sarkovskii ordering as the bifurcation parameter a is increased.[15] Chaos appears for a $\gtrsim$ 3.5699..., above which various windows exist among the chaotic regions. May emphasized the importance of the map (1.2.3), which has been proven by recent progresses.

Since the divergence $(\partial \dot{X}/\partial X + \partial \dot{Y}/\partial Y + \partial \dot{Z}/\partial Z) = -(\sigma + b + 1)$ is negative for the Eq. (1.2.1), the volume in phase space shrinks to zero. This does not, however, mean that the attractor is confined in a two-dimensional surface. As a matter of fact, it is mathematically proven that chaos is impossible for a flow in a two-dimensional surface. The Lorenz attractor has an infinite sheet structure (Cantor structure) and has a dimension larger than two. Patterns with noninteger dimensions have been intensively and extensively studied by B.B. Mandelbrot,[16] who called them "fractals". They are characterized by self-similarity (see Fig. 1.4).

Lorenz attractor has this type of fractal structure. In this sense, one-dimensional property of the Lorenz plot (Fig. 1.3) is not perfect. This feature can be clearly seen in the Hénon map[17]

$$x_{n+1} = ax_n(1 - x_n) + y_n$$

$$y_{n+1} = bx_n \tag{1.2.4}$$

which is an invertible map and can be considered as a Poincaré map of some flow system. The Jacobian of the map (1.2.4) is given by $-b$. If $b = 0$, the dissipation is so large that the map (1.2.4) reduces to the one-dimensional map (1.2.3). For $b \neq 0$, however, the dimension of the attractor can be larger than one. Fractal structure of the attractor can be seen in Fig. 1.5. The one-dimensional structure of the Lorenz plot is obtained only if this Cantor structure is neglected.

8

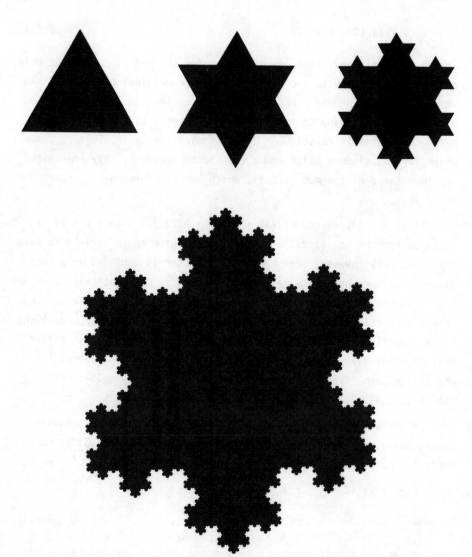

Triadic Koch island or snowflake 𝒦. Original construction
by Helge Von Koch (coastline dimenseion D = log 4/log 3 ~ 1.2618).

Fig. 1.4 Construction of "fractals"
(cited from Ref. (16)).

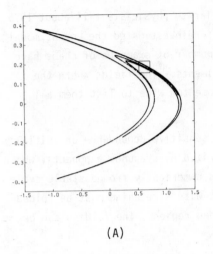

(A)

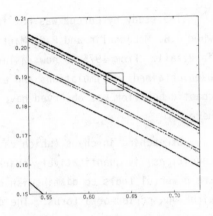

Enlargement of the squared region of Figure A. The number of computed points is increased to $n = 10^5$.

(B)

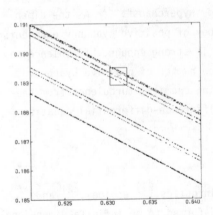

Enlargement of the squared region of Figure B: $n = 10^6$.

(C)

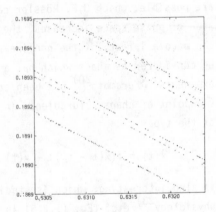

Enlargement of the squared region of Figure C: $n = 5 \times 10^6$.

(D)

Fig. 1.5 Attractor of the Hénon map (cited from Ref. (17)). The transversal axis denotes $(x-1/2)/(a/4-1/2)$ and the longitudinal axis denotes $(y-1/2)/(a/4-1/2)$. ($a = 1 + \sqrt{6.6}$ and $b = 0.3$).

The study by Lorenz was not familiar to physicists till about 1975, when J.B. McLaughlin and P.C. Martin[18] reinvestigated the Lorenz model in detail. From 1975 to now, a large number of examples of chaos have been obtained in simulations and experiments. The fields where the chaotic behavior was observed have become too wide to list them all here.

Stretching in chaos (which causes sensitive dependence on initial conditions) is quantitatively characterized by Lyapunov exponents, which are powerful tools to distinguish chaos numerically from a stable cycle with long period or a torus. The orbit with stretching must be folded sometimes in order to remain in a bounded region. The folding can be typically seen in the map in Fig. 1.3.

Lorenz's strange attractor has only one positive Lyapunov exponent (i.e., only one direction with stretching). For higher-dimensional flows (dimension ≥ 4), attractors with many positive Lyapunov exponents are possible, which O.E. Rössler called "hyperchaos".[19] As the dimension of phase space increases, the number of positive Lyapunov exponents can become larger, if the nonlinearity is strong enough. Turbulence may be considered as chaos which has a large number of positive Lyapunov exponents. Kuramoto[20] has been studying chemical turbulence from the viewpoint of chaos. Turbulence in a difference-differential equation of the type

$$\gamma^{-1}\dot{x}(t) = f(x(t - t_R)) - x(t) \qquad (1.2.5)$$

is also interesting, which is studied in optics,[21] ecology,[22] and physiology[23] etc. (Eq. (1.2.5) is transformed to an infinite-dimensional mapping). Farmer has studied the distribution of positive Lyapunov exponents for this type of equation.[24] Of course, turbulence in fluids is an important and fascinating phenomenon which has to be attacked from the viewpoint of chaos.[25]

Since the unpredictability lies in the deterministic chaos, statistical methods must be introduced. Attempts to construct a statistical mechanics have been carried out by Ruelle and Bowen,[26]

Takahashi and Oono,[27] Kai and Tomita[28] and others, though many problems are still left to be solved in the future.

Theoretical studies which characterize the property of chaos itself, still remain unsatisfactory. On the other hand, studies on the mechanism of the onset of chaos have made a large progress in these few years, which will be described in the following section.

§1.3 Onset of Chaos

How does the chaos (or turbulence) appear? It has been an unresolved and fascinating question for many years. Landau proposed the following picture for the onset of turbulence;[29]

    limit cycle - (Hopf bifurcation) → 2-torus

    - (Hopf bifurcation) → 3-torus → ....

    → ∞-torus = turbulence,

where n-torus is a quasiperiodic state with n incommensurate frequencies. This picture of turbulence needs an infinite-dimensional dynamical system. Thus, it seemed very difficult to study the onset of turbulence.

In 1971, Ruelle and Takens wrote a remarkable paper.[30] They showed that the 3-torus is unstable against some structural perturbations and strange attractors exist at any neighborhood of a 3-torus. Thus, their picture is shown by

    limit cycle - (Hopf bifurcation) → (2-torus)

    - [with various lockings] - (Hopf bifurcation)

    → chaos.

According to this picture, a high-dimensional dynamical system is not necessary for the study of the onset of chaos. Thus, their paper lifts the curtain on the chaos in low-dimensional dynamical systems, though the theory has been, in some sense, overevaluated (see Chap. 6).

Before we investigate the transition from torus to chaos, we review briefly other routes to chaos, which are more fundamental. The most "famous" route is period-doubling bifurcations to chaos. As the bifurcation parameter A is changed, the period-doubling of a limit cycle occurs at $A = A_n$ successively in a variety of systems. Since the number of relevant dimension is one in this problem, one-dimensional mapping $x_{n+1} = f(x_n)$ is useful and powerful, as was shown for example by May,[14] Grossman and Thomae,[31] Feigenbaum,[32] and Tresser and Coullet.[33] The critical phenomena and renormalization group approach has been intensively studied by Feigenbaum which is well-known as Feigenbaum's theory.[32]

Feigenbaum found for a wide class of one-dimensional mappings that

$$A_\infty - A_n \propto \delta^{-n} , \qquad (1.3.1)$$

where A_∞ is the value of the accumulation of period-doubling and the onset of chaos. The constant δ is now known as a Feigenbaum's constant. This universal constant depends only on the value z, which is the exponent at the top of the map $f(x)$;

$$f(x) \sim \text{const.} - A(x - x_0)^z \qquad (1.3.2)$$

(for $z = 2$, $\delta = 4.6692016091029909...$).

The critical phenomena including other critical exponents are well explained by the renormalization group theory, which was introduced to the theory of chaos by Feigenbaum. Universality of the critical phenomena of the period-doubling bifurcations is shown[34] for the class of mappings which are single-humped and satisfy the Schwarzian condition (i.e., $f'(x)f'''(x)/3! - (f''(x)/2)^2 < 0$). The abnormality of period-doubling bifurcations can occur if the Schwarzian condition is broken, which was investigated by I. Tsuda[35] in connection with the experiments on BZ reactions.

A lot of experiments have been performed, which confirm Feigenbaum's theory. Especially, experiments on the Rayleigh-Bénard convection have been extensively performed,[36] which show the period-doubling up to the 2^5-period and give the value $\delta \sim (4 \sim 5)$ (see Fig. 1.6). The window structure of the logistic map was also observed in the Rayleigh-Bénard experiment by A. Libchaber.[37] If the effect of noise is taken into account, only a finite number of the doublings can be observed. The

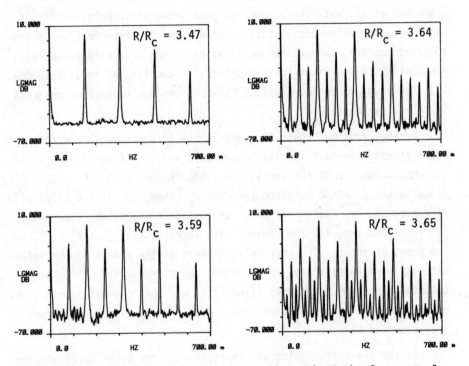

The Fourier spectrum. Arrows indicate the peak at the frequency f_1.

Fig. 1.6 Power spectrum of the temperature variation in a Rayleigh-Bénard experiment in mercury. The experimental cell has an aspect ratio $\Gamma = 4$ and contains four convective rolls. A DC magnetic field of 270 G is applied along the roll axis. The measured Feigenbaum constant is $\delta = 4.4$. (R is a Rayleigh number and R_c is the number for the onset of convection.) (cited from Ref. 36) A. Libchaber et al., J. de Phys. L43 (1982).

scaling relation including the noisy effect was first studied by J.P. Crutchfield and B.A. Huberman[38], which was elegantly formulated by Schraiman et al. with the use of the path integral formulation.[39,40] There are also other possibilities which cause the interruption of the doubling cascade, such as the multibasins in two-dimensional mappings,[41] abnormality in one-dimensional mappings,[34] doubling of torus (see Chap. 5). The study of period-doubling bifurcations has made great progress in these few years.

Another familiar route to chaos is the intermittency. In this case, chaos appears immediately after a periodic solution loses its stability. The chaos consists of two parts, i.e., the laminar part (the orbits stay close to the periodic solution for $\varepsilon < \varepsilon_c$ where ε is the bifurcation parameter and ε_c is the value of the onset of chaos) and the burst. The residence time for the laminar part increases as $(\varepsilon - \varepsilon_c)^{-1/2}$ near the onset of chaos. The critical phenomena of the transition by intermittency were investigated by Y. Pomeau and P. Manneville.[42] The critical exponents are simple (like 1/2), and the renormalization group equation has an exact solution, which was given by Hirsch, Huberman, Scalapino and Nauenberg.[43]

At the transition point of intermittency, the power spectrum shows an f^{-1} behavior.[44,45] Recently Y. Aizawa and his coworkers have succeeded in extracting the nonstationary features of chaos at the intermittency,[46] in connection with the Pareto-Zipf's law.

So far, we have considered a system with one attractor. A dynamical system, however, can have more than one attractor and the basin of attraction is split into many parts (multibasin phenomena).[47] The basin boundary can be fractal in some systems, which was investigated by Mandelbrot,[16] and Grebogi et al.[48] The critical phenomenon when one chaotic attractor disappears (which is analogous to the first order phase transition) was investigated by C. Grebogi et al., who called this transition as "crisis".[49,50] The basin structure itself can become fractal as was noted by Fatou and Mandelbrot,[51] which was investigated by Takesue and the author[52] including the effect of noise on the structure.

So far, we have omitted one important route to chaos, i.e., the transition from torus to chaos, which is the main theme of the present book.

§1.4 Transition from Torus to Chaos Accompanied by Lockings — Outline of the Book

Transition from torus to chaos accompanied by frequency lockings is an important route to chaos, which has been extensively and intensively studied in recent few years. Before 1982, there were only few theoretical (or computation-physical) works on the transition from torus to chaos in dissipative systems.

Studies on the stability of tori in a Hamiltonian system, on the other hand, have a long history.[53] Poincaré had already known the complicated behavior in Hamiltonian systems. Stability of the torus is established in the celebrated KAM (Kolmogorov-Arnold-Moser) theory.[54] As the perturbation is increased, the KAM tori collapse successively and stochastic (chaotic) region in a phase space increases.[55] Critical phenomena at the collapse of the last KAM torus were studied by J.M. Greene,[56] L.P. Kadanoff and S.J. Shenker.[57]

Stability of a torus in dissipative systems was first studied by Ruelle and Takens in 1971,[30] which gave a large impact on experimental physicists, who began to work on the Rayleigh-Bénard instability or the Taylor-Couette instability. Simulations were few, however, except the one for the Navier-Stokes equation truncated by 48 modes by H. Yahata.[58] A large progress has been performed in these few years.

In the present book, collapse of tori is investigated using various types of mappings. When the Poincaré map is taken, we have a d-dimensional mapping $(d \geq 2)$ (see Fig. 1.7). If our interest is restricted only to the phase motion of tori, however, one-dimensional mapping $\theta_{n+1} = g(\theta_n)$ (mod 1) is useful. In Chap. 2, we study the phase

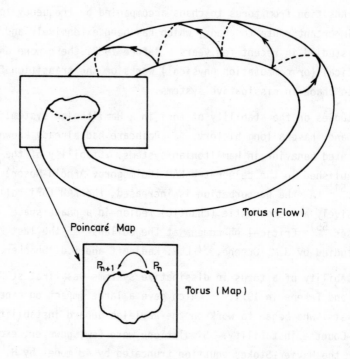

Torus (Flow)

Poincaré Map

P_{n+1} P_n

Torus (Map)

Fig. 1.7 Schematic representation of the construction of
a mapping from a torus motion.

motion of tori using the one-dimensional map $\theta_{n+1} = \theta_n + A \sin (2\pi\theta_n) + D$ (mod 1). The structure of lockings is briefly shown in §2.2. Among the various lockings, "period-adding" sequence is chosen to study the similarity structure of lockings. Phenomenological theory based on the existence of a fixed point function is constructed, which explains well the numerical observation on similarity. Applications of the theory to the sequence of windows are also given. Chaotic trajectories which appear after the collapse of tori are characterized by the disordering property introduced in §2.9. Some reviews on recent renormalization group study and experimental results are given in §2.7, §2.8 and §2.10.

In Chap. 3, collapse of tori in two-dimensional mappings are investigated using the coupled-logistic map. Period-adding sequence is again chosen to give the scaling relations. The numerical results are well explained by the theory of Chap. 2. In a system which has some symmetry like the coupled-logistic map, however, locking with symmetry breaking can occur. The discovery of this phenomena is also reported. When the symmetry is broken, two types of attractors coexist. The structure of basins of attraction ("self-similar stripe structure") is also studied, though the detailed analysis on basin structures will be reported elsewhere.[52]

Instability of tori due to the amplitude motion is also important. In Chap. 4, oscillation of tori along the amplitude direction is investigated. First, the oscillation of tori is related to the oscillation of an unstable manifold of a periodic saddle. A simple two-dimensional mapping ("delayed-logistic map") is introduced and investigated. In usual two-dimensional mappings, however, oscillation is masked by lockings. In §4.3, "modulation map" is introduced to study the oscillatory instability in more detail. The torus loses its differentiability at the critical point, above which chaos appears. Fractal dimension of the torus at the critical point is numerically calculated on the basis of a "functional map", which is obtained from the equation for the invariant curve.

In higher-dimensional systems, torus itself can show period-doubling bifurcations. In Chap. 5, the doubling of torus is investigated in detail. Discovery of doubling of torus is given in §5.1. These examples show only a finite number of doublings before chaos appears, which is studied using a coupled logistic and torus map. The mechanism of the interruption of doubling is also investigated in connection with the oscillation of torus given in Chap. 4.

If the torus undergoes a Hopf bifurcation, a three-torus can appear. Though Ruelle and Takens showed the structural instability of three-tori and the emergence of strange attractor,[30] structural instability of a state is not sufficient to conclude that the state is not physically observable. In Chap. 6, stability of a three-torus is investigated using two- or four-dimensional coupled maps. A three-torus stably exists for a weak coupling, which becomes feasible to lock into a torus or a cycle as the coupling is increased. Transition to chaos occurs only via a locking into a cycle. Thus, the mechanism of the onset of chaos by Ruelle and Takens cannot be observed in our models. Lockings into tori form a "double devil's staircase", which is studied by the modulated circle map given in §6.3. Detailed results on coupled circle maps are shown in §6.4.

In Chap. 7, a new topic is presented, that is, coupled map lattice (CML) is introduced to study a "field theory of chaos". In CML, one-dimensional maps are coupled on a lattice. The objectives of the CML research are as follows:

  i)   to study a system with a large number of excited modes;

 ii)   characterization of spatial patterns such as kinks;

iii)   construction of a statistical mechanics for a system with spatiotemporal complexity;

 iv)   to study the onset and development of chaos as a bifurcation parameter is changed;

  v)   modelling of a system with a spatial bifurcation such as an open flow.

Various numerical results for the spatiotemporal patterns in CML are shown. Relations with cellular automata are also discussed.

Chapter 8 is devoted to summary, future problems, and discussions. What chaos has brought about and will bring about in science is discussed in §8.2, where a limit of our science, "nontrivial" infinity and non-stationarity in chaos, impact of chaos on various fields such as biology, social science, and orthodox physics are considered. Some considerations on coupled map lattices and cellular automata are given in §8.3, in connection with the field theory of chaos, turbulence, and neuron networks.

References

1. T.S. Kuhn, The Structure of Scientific Revolutions, (Univ. of Chicago Press, Chicago, 1962).

2. Recently I. Prigogine has pointed out the similar picture in a little different context; I. Prigogine, From Being to Becoming, (W.H. Freeman and Company, 1980).

3. See for the recent growth of computer physics, Special Issue of Phys. Today 36 (1983) pp. 24-63 and p. 128, by D. Hamann, M. Creutz, J.E. Hirsch and D.J. Scalapino, I. Cohen and J. Killeen, and K.G. Wilson.

4. M. Born, Physicalische Blätter 11 (1955) 49.

5. W. Heisenberg, Phys. Today 20 (1967) 27.

6. S. Tomonaga, What is Physics? (Iwanami, Tokyo, 1980) in Japanese.

7. Developments in this field up to 1978 can be seen in Suppl. Prog. Theor. Phys. 64 (1978).

8. Recent progress can be seen in P. Glansdorff and I. Prigogine, Thermodynamics of Structure, Stability, and Fluctuations (Wiley-Interscience, 1971) and G. Nicolis and I. Prigogine, Self-Organization in Nonequilibrium Systems, (Wiley-Interscience, 1977).

9. E. Schrödinger, What is life? (Cambridge Univ. Press, N.Y., 1945).

10. Revolt against reductionism was performed by L. von Bertalanffy, General System Theory (George Braziller, N.Y., 1968).

11. H. Haken, Synergetics, An Introduction (Springer, 1977), Advanced Synergetics (Springer, 1983).

12. Many examples can be seen in Springer Series in Synergetics.

13. E.N. Lorenz, J. Atmos. Sci. 20 (1963) 130.

14. R. May, Nature 26 (1976) 459.

15. A. Sarkovski, Ukr. Mat. Zh. 16 (1964) 61.

16. B.B. Mandelbrot, The Fractal Geometry of Nature (Freeman, 1982).

17. M. Hénon, Comm. Math. Phys. 50 (1976) 69.

18. J.B. McLaughlin and P.C. Martin, Phys. Rev. A12 (1975) 186.

19. O.E. Rössler, Phys. Lett. 71A (1979) 155.

20. Y. Kuramoto, Physica 106A (1981) 128.

21. K. Ikeda, H. Daido and O. Akimoto, Phys. Rev. Lett. 45 (1980) 709.

22. R. May, Annals of N.Y. Acad. Sci. 357 (1980) 267.

23. M.C. Mackey and L. Glass, Science 197 (1977) 287.

24. J.D. Farmer, Physica 4D (1982) 366.

25. Some attempts towards this direction were seen in the IUTAM symposium on Turbulence and Chaotic Phenomena in Fluids (Kyoto, 1983).

26. R. Bowen and D. Ruelle, Inventiones Math. 29 (1975) 181.

27. Y. Oono and Y. Takahashi, Prog. Theor. Phys. 63 (1980) 1804.

28. T. Kai and K. Tomita, Prog. Theor. Phys. 64 (1980) 1532.

29. L.D. Landau, and E.M. Lifshitz, Fluid Mechanics (Pergamon, London, 1959) Chap. III.

30. D. Ruelle and F. Takens, Comm. Math. Phys. 20 (1971) 167.

31. S. Grossman and S. Thomae, Z. Naturforschung 32A (1977) 1353.

32. M.J. Feigenbaum, J. Stat. Phys. 19 (1978) 25; 21 (1979) 669; Phys. Lett. 74A (1979) 375; Comm. Math. Phys. 77 (1980) 65.

33. C. Tresser and P. Coullet, CR. Acad. Sci. 287A (1978) 577.

34. P. Collet and J.P. Eckman, Iterated Maps on the Intervals as Dynamical Systems (Birkhäuser, 1980).

35. I. Tsuda, Prog. Theor. Phys. 66 (1981) 1985.

36. J. Maurer and A. Libchaber, J. Phys. L40 (1979) L-419; A. Libchaber, C. Laroche and S. Fauve, J. de Phys. L43 (1982) L-211; M. Giglio, S. Musazzi, and U. Perini, Phys. Rev. Lett. 47 (1981) 243.

37. A. Libchaber, S. Fauve and C. Laroche, Physica 7D (1983) 73-84.

38. J.P. Crutchfield and B.A. Huberman, Phys. Lett. 74A (1980) 407.

39. B. Schraiman, C.E. Wayne and P.C. Martin, Phys. Rev. Lett. 46 (1981) 935.

40. See also J.P. Crutchfield, M. Nauenberg and J. Rudnick; Phys. Rev. Lett. 46 (1981) 933.

41. A. Arnéodo, P. Coullet, C. Tresser, A. Libchaber, J. Maurer and D. d'Humières, Physica 6D (1983) 385.

42. Y. Pomeau and P. Manneville, Comm. Math. Phys. 74 (1980) 189.

43. J.E. Hirsch, B.A. Huberman and D.J. Scalapino, Phys. Rev. A25 (1982) 519; J.E. Hirsch, M. Nauenberg and D.J. Scalapino, Phys. Lett. 87A (1982) 391.

44. P. Manneville, J. de Phys. 41 (1980) 1235.

45. I. Procaccia and H. Schuster, Phys. Rev. A28 (1983) 1210.

46. Y. Aizawa, Prog. Theor. Phys. 70 (1983) 1249, Y. Aizawa and T. Kohyama, Prog. Theor. Phys. 71 (1984) 847; Y. Takahashi, private communication.

47. K. Tomita and H. Daido, Phys. Lett. 79A (1980) 133; C. Tresser, P. Coullet and A. Arnéodo, J. de Phys. Lett. 41 (1980) L243.

48. C. Grebogi, E. Ott and J.A. Yorke, Phys. Rev. Lett. 50 (1983) 935.

49. This problem was first treated by H. Daido; Prog. Theor. Phys. 63 (1980) 1190.

50. C. Grebogi, E. Ott and J.A. Yorke, Phys. Rev. Lett. 48 (1982) 1507.

51. See Ref. 16.

52. S. Takesue and K. Kaneko, Prog. Theor. Phys. 71 (1984) 35.

53. See e.g. V.I. Arnold and A. Avez, Problèmes Ergodiques de la Mécanique Classique (Gauther-Villars 1967) (Translation (Japanese) by K. Yosida).

54. See e.g., J. Moser in Stable and Random Motions in Dynamical Systems, (Princeton Univ. Press, Princeton, 1973).

55. See e.g., B.V. Chirikov, Phys. Rep. 52 (1979) 263.

22

56. J.M. Greene, J. Math. Phys., 9 (1968) 760, 20 (1979) 1183.

57. S.J. Shenker and L.P. Kadanoff, J. Stat. Phys. 27 (1982) 631; L.P.
     Kadanoff, in Melting, Localization, and Chaos (ed. R.K. Kalia and
     P. Vashishta (Elsevier, N.Y., 1982); see also, D.F. Escande and F.
     Doveil, J. Stat. Phys. 26 (1981) 257.

58. H. Yahata, Prog. Theor. Phys. 64 (1980) 782.

Chapter 2

INSTABILITY OF PHASE MOTION OF TORI

Dew Drop
by *M. C. Escher*

§2.1 Introduction

When our interest is restricted only to the phase motion of tori, one-dimensional map

$$\theta_{n+1} = f(\theta_n) \quad (\text{mod } 1) \; ;$$

$$f(\theta) \text{ is continuous with } f(0) = f(1) \quad (\text{mod } 1) \qquad (2.1.1)$$

is very useful and powerful. As a simple and typical case, we consider the map

$$\theta_{n+1} = f(\theta_n) = \theta_n + A \sin (2\pi\theta_n) + D \quad (\text{mod } 1) \; . \qquad (2.1.2)$$

For $|A| < 1/(2\pi)$, the map is invertible and the attractor is torus or locking. The torus and locking (cycle) is numerically distinguished by

i) Period (torus → ∞, locking → finite),

ii) Lyapunov exponents (torus → 0, locking → < 0),

iii) rotation number (torus → irrational, locking → rational), where the rotation number ρ is defined by

$$\rho = \lim_{n \to \infty} (f^n(\theta) - \theta)/n \; . \qquad (2.1.3)$$

A cycle exists stably in an interval in a parameter space, where ρ takes a constant value, which is the reason why it is called "locking". The rotation number forms a devil's staircase as a function of the bifurcation parameter.

For $|A| > 1/(2\pi)$, the map is noninvertible and the attractor is chaos or window (cycle). In the region $|A| > 1/(2\pi)$, various routes to chaos, i.e., period-doubling, intermittency, and crisis are observed. It is also important to note that the multibasin phenomena can happen for $|A| > 1/(2\pi)$.

There arise a lot of questions about the circle map (2.1.2):

i) What is the mechanism of lockings?

ii) Are there any similarities (and scaling relations) between the lockings?

iii) What is the structure of windows? Are there any similarities between windows?

iv) What is the critical phenomenon at the collapse of tori with a given irrational rotation number?

v) How are the chaotic trajectories characterized? etc.

In this chapter some efforts to answer these questions are described. In §2.2, structure of lockings of the map (2.1.2) is roughly described. Among the various lockings, a "period-adding sequence" is chosen to study the various scalings and the similarity. In §2.3, numerical discoveries on the critical phenomena of the period-adding sequence are reported, while the theory based on the fixed point function is described in §2.4, which explains the numerical results well. The classification of the sequence into three categories according to this theory is shown in §2.5. A period-adding sequence as windows also appears in various systems, which is discussed in §2.6, though it may not have direct relations to the transition from torus to chaos.

At $|A| = 1/(2\pi)$, the map loses its invertibility and torus disappears. Critical phenomena at the collapse of tori are reviewed in §2.7. In §2.8, global properties of the devil's staircase are studied. The study on the properties of chaotic trajectories for $|A| > 1/(2\pi)$ is still premature. In §2.9, the "disordering" property of a chaotic trajectory is investigated to characterize the chaos. Discussions are given in §2.10, where experimental results on the transition from torus to chaos are briefly reviewed.

§2.2 Structure of Lockings

For $A < 1/(2\pi)$, the attractor of the map (2.1.2) consists of torus and locking. The measure of the locking (in the parameter space) is zero at $A = 0$ and it increases as A is increased. The rough phase diagram for $A < 1/(2\pi)$ is given in Fig. 2.1. Between the lockings with q/p and s/r, (p,q and r,s are relatively prime integers respectively; "the locking with q/p" means the locking with the rotation number q/p and with the period p) there is a locking with

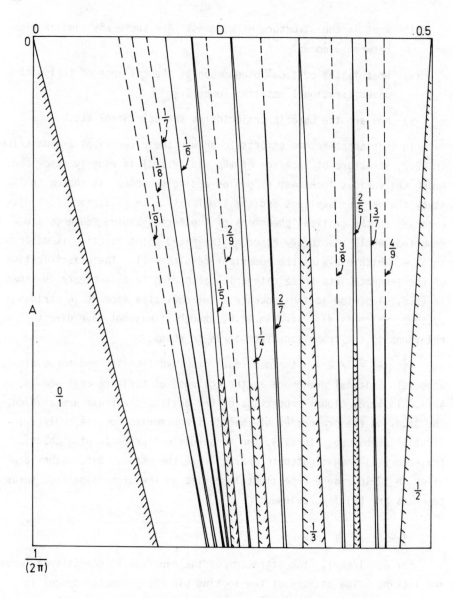

Fig. 2.1 Rough phase diagram of the map (2.1.2) for
A < 1/(2π). Only the cycles with periods
less than 10 are shown (blank regions are
tori or cycles with periods larger than 9).
The number q/p denotes the rotation
number and p shows the period.

(q + s)/(p + r). Thus, the locking is constructed from the Farey series (see Fig. 2.2 schematically). Furthermore, it is known that the rotation number as a function of the bifurcation parameter (e.g., D) forms an incomplete devil's staircase, i.e., the measure of torus is finite (see Figs. 2.3a)-2.3c)). The increase of the measure of the lockings as the increase of A is typically seen from these figures.

How does the locking from torus occur? The function $y = f^p(x)$ is tangential to $y = x$ at p points at which the locking with q/p occurs (see for example Fig. 2.4 for $y = f^5(x)$ at the onset of a 5-cycle). Thus, the locking phenomenon from torus has a common feature with the intermittency theory in the sense that the tangent bifurcation is essential to both phenomena.

§2.3 Similarity of the Period-Adding Sequences of Lockings (Numerical Results)

In this section we study the similarity of a period-adding sequence using the map (2.1.2). The "period-adding" sequence means the sequence of the lockings with $(qn + s)/(pn + r)$ $(n = 1,2,...)$. The reason for this choice is that

i)   this sequence is feasible to be observed since it has a large stability;

ii) it makes clear the mechanism of lockings; and

iii) it reveals the global property of lockings through various scalings and similarity structures.

We define A_n (A_n^f) by the parameter of A at which the locking with $(qn + s)/(pn + r)$ appears (disappears) $(A_n^f > A_n)$. The numerical simulations were performed on the map (2.1.2), which show

$$A_\infty - A_n \propto n^{-2} \qquad (2.3.1)$$

and

$$\Delta A_n \equiv A_n^f - A_n \propto n^{-3} . \qquad (2.3.2)$$

Here, we take A as a bifurcation parameter. The similarity structure

28

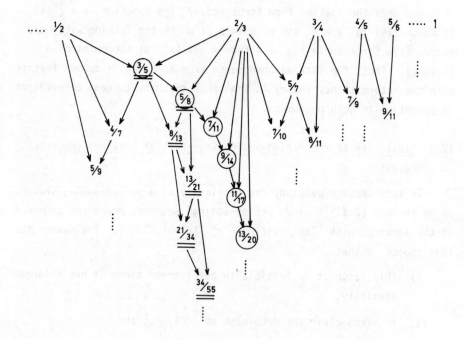

Fig. 2.2 Construction of lockings (Farey series)
○ → Period-adding sequence;
═ → Fibonacci sequence

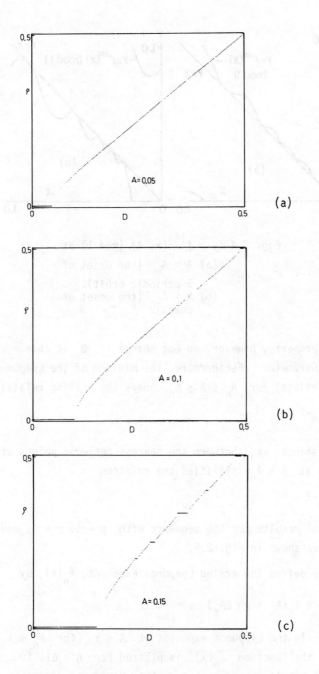

Fig. 2.3 Rotation number ρ as a function of D
          for the map (2.1.2) (Devil's staircase).

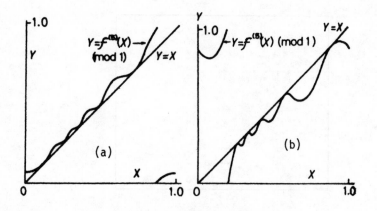

Fig. 2.4 $y = f^{(5)}(x; A)$ (mod 1) at
(a) $A = A_\infty$ (the onset of
5-periodic orbit).
(b) $A = A_C$ (the onset of
chaos).

and scaling property, however, do not change if D is chosen as a
bifurcation parameter. Furthermore, the minimum of the Lyapunov exponent (if it exists) for $A_n \leq A \leq A_n^f$ obeys the scaling relation

$$\lambda_n^{min} \propto n^{-1} , \qquad (2.3.3)$$

while the distance Δx_n between the nearest periodic points of the
(pn+r)-cycle at $A = A_n$ satisfies the relation

$$\Delta x_n \propto n^{-2} . \qquad (2.3.4)$$

Some numerical results for the sequence with $p = 5$, $r = 4$, and $q = s = 1$
(D = 0.25) are shown in Fig. 2.5.

Next, we define the scaled Lyapunov exponent $\tilde{\lambda}_n(x)$ by

$$\tilde{\lambda}_n(x) \equiv n \lambda_n(A_n + x \cdot \Delta A_n) , \qquad (2.3.5)$$

where $\tilde{\lambda}_n(a)$ is the Lyapunov exponent at $A = a$ (for $A_n < a < A_n^f$).
In Fig. 2.6, the function $\tilde{\lambda}_n(x)$ is plotted for $n = 61$, 190, and 295.
The numerical result, as is seen from this figure, suggests that $\tilde{\lambda}_n(x)$

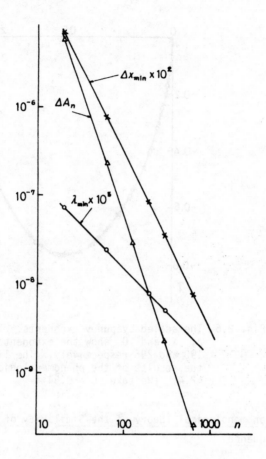

Fig. 2.5 Various scaling properties for the
(5n+4)-sequence at D = 0.25. The
quantities $\log(\Delta A_n)$, $\log(\Delta x_n)$ and
$\log \lambda_n^{min}$ are depicted versus log n.

approaches a fixed point function $\tilde{\lambda}(x)$ as n goes to infinity. We
note that $\tilde{\lambda}(x) \propto -x^{1/2}$ near $x \approx 0$ and $\tilde{\lambda}(x) \propto -(1-x)^{1/2}$ near
$x \approx 1$. The asymmetry about $x \approx 0.5$ is not found for the function $\tilde{\lambda}(x)$
corresponding to the sequence with p = 5, r = 4 and q = s = 1 (D =
0.25). We also have to note that the function $\tilde{\lambda}(x)$ is <u>not</u> universal.
The functions $\tilde{\lambda}(x)$ corresponding to other sequences have different
shapes from Fig. 2.6.

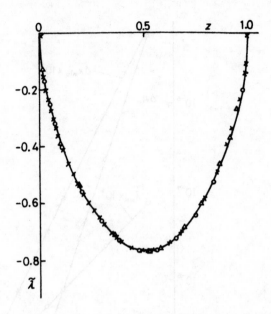

Fig. 2.6 The scaled Lyapunov exponents. The notations
Δ, X and 0 show the exponents for n = 61,
190 and 295 respectively. The line denotes
the results of the phenomenological theory of
§2.4. (We take C̃ = 0.54)

§2.4 <u>Phenomenological Theory of the Similarity of the Period-Adding</u>
<u>Sequence</u>

The observations in §2.3 can be understood through the following
considerations.

i) The locking occurs via tangent bifurcations. Thus, the argument similar to Pomeau and Manneville's intermittency theory[1] is valid.
The function $y = f^p(x)$ at $A = A_\infty$ is tangential to $y = x$ at
$x = x_1^*, x_2^*, \ldots x_p^*$. The function at $A = A_n$, therefore, is expanded
around the periodic points $\{x_\nu^*\}$ $(\nu = 1, 2, \ldots, p)$ by

$$f^p(x) \approx x + a_\nu (x - x_\nu^*)^2 + \varepsilon_n \pmod 1 \qquad (2.4.1)$$

where ε_n is proportional to $(A_\infty - A_n)$. From Eq. (2.4.1), we have

$$(\text{rot. no. at } A_n) - (\text{rot. no. at } A_\infty) \propto \sqrt{\varepsilon_n} \quad , \qquad (2.4.2)$$

since the number of the orbit points close to x^*_ν is $O(1/\sqrt{\varepsilon_n})$ while the number of the points far from x^* is $O(1)$. On the other hand, the difference between the rotation numbers at $A = A_n$ and $A = A_\infty$ is given by

$$(qn + s)/(pn + r) - q/p \propto (ps - rq)/(p^2n) \quad \text{as} \quad n \to \infty \quad . \qquad (2.4.3)$$

From Eqs. (2.4.2) and (2.4.3) we have $\varepsilon_n \propto n^{-2}$. Thus, the scaling relation (2.3.1) is derived.

Since Δx_n is the distance between the nearest periodic points close to x^*_ν, it is given approximately by $f^p(x^*_\nu; A_\infty) - x^*_\nu$. The expansion (2.4.1), therefore, gives the scaling relation (2.3.4), because ε_n is proportional to n^{-2} for $n \to \infty$.

ii) (The assumption of the existence of the fixed point function.)[2)]

We define the following scaled function $(0 \leq x \leq 1)$ by

$$f_n(x) = \alpha_n^{-1}(f^{pn+r}(\alpha_n x + x_0^n; A_n) - x_0^n) \quad , \qquad (2.4.4)$$

where x_0^n and x_1^n are the nearest pair of periodic points of $(pn+r)$-cycle and α_n is a scaling factor so that $\alpha_n^{-1}(x_1^n - x_0^n) = 1$ (see Fig. 2.7 schematically). We make an assumption that

$$f_n(x) \to f^*(x) \quad \text{as} \quad n \to \infty \quad . \qquad (2.4.5)$$

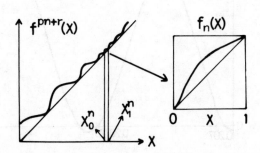

Fig. 2.7 Schematic representation of the construction of $f_n(x)$.

34

The assumption seems to be valid from numerical results. The functions $f^{19}(x) - x$ (p = 5, r = 4, and q = s = 1) at $A \approx A_3$ and $A \approx A_3^f$ are given in Fig. 2.8, while $f^{309}(x) - x$ (p = 5, r = 4, and q = s = 1 and n = 61) is given in Fig. 2.9. These figures are useful to make a phenomenological theory of the similarity.

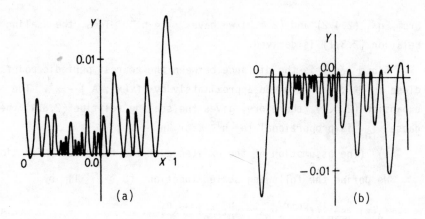

(a)

(b)

Fig. 2.8 (a) $f^{19}(x) - x$ at A = 0.14813 and
(b) at A = 0.14842.

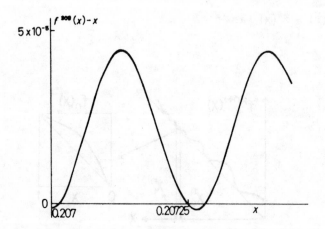

Fig. 2.9 $f^{309}(x) - x$ at A = 0.156673819 for
0.2070 < x < 0.2074.

iii) As the bifurcation parameter A is changed from A_n, the function $f^{pn+r}(x; A)$ also changes. If we confine ourselves only to the smallest structure of $f^{pn+r}(x)$ (i.e., $x_0^n \leq x \leq x_1^n$), the main change of the form $f^{pn+r}(x; A)$ is the addition of constants. Thus, we have approximately

$$f^{pn+r}(x; A) \simeq \alpha_n f^*(\alpha_n^{-1}(x - x_0^n)) + x_0^n - \frac{A - A_n}{\gamma_n} \qquad (2.4.6)$$

for large n. The new scaling factor γ_n is given by $(\partial f^{pn+r}(x_0^n)/\partial A)^{-1}_{A=A_n}$, which is proportional to n^{-1} from numerical results.

The scaling $\gamma_n \propto n^{-1}$ can be explained as follows: Using the chain rule we have

$$f^{pn+r}(x; A_n + \delta A) - f^{pn+r}(x; A_n) = \frac{\partial f^{pn+r}(x; A_n)}{\partial A} \delta A$$

$$= \delta A \left\{ \frac{\partial f^p(x_{p(n-1)+r}; A)}{\partial A} \bigg|_{A_n} + \frac{\partial f^p(x_{p(n-2)+r}; A)}{\partial A} \bigg|_{A_n} \frac{\partial f^p(x_{p(n-2)+r}; A_n)}{\partial x_{p(n-2)+r}} \right.$$

$$+ \ldots + \frac{\partial f^p(x_{p+r}; A)}{\partial A} \bigg|_{A_n} \frac{\partial f^p(x_{p(n-2)+r}; A_n)}{\partial x_{p(n-2)+r}} \times \frac{\partial f^p(x_{p(n-3)+r}; A_n)}{\partial x_{p(n-3)+r}}$$

$$\left. \times \ldots \times \frac{\partial f^p(x_{p+r}; A_n)}{\partial x_{p+r}} \right\} , \qquad (2.4.7)$$

where x_k denotes $f^k(x)$. Since each term is independent of n as n gets larger and the number of terms is n, we have

$$f^{pn+r}(x; A_n + \delta A) - f^{pn+r}(x; A_n) \propto n \cdot \delta A , \qquad (2.4.8)$$

which means $\gamma_n \propto n^{-1}$. Though γ_n can depend on x in principle, we can approximate it as a constant, since x goes to x_0^n as n gets larger and α_n goes to 0.

iv) Now, we consider the Lyapunov exponents. We can write

$$\lambda_n = \frac{1}{pn + r} \log |f^{(pn+r)'}(x_n^*)| \quad , \tag{2.4.9}$$

where x_n^* is an arbitrary fixed point of the map $x \rightarrow f^{pn+r}(x)$. We choose such a fixed point x_n^* that satisfies $x_0^n < x_n^* < x_1^n$. Using Eq. (2.4.6), we have

$$\lambda_n = \frac{1}{pn + r} \log \left\{ |1 + g^{*'}(z_n^*)| \right\} \quad , \tag{2.4.10}$$

where

$$g^*(x) = f^*(x) - x \tag{2.4.11}$$

and $z_n^* = \alpha_n^{-1}(x_n^* - x_0^n)$. The fixed point x_n^* is given by

$$x_n^* + \alpha_n g^*(\alpha_n^{-1}(x_n^* - x)) - \frac{1}{\gamma_n}(A - A_n) = x_n^* \quad . \tag{2.4.12}$$

Thus we have $z_n^* = g^{*(-1)}((A - A_n)/\alpha_n \gamma_n)$, where $g^{*(-1)}$ is an inverse function of $g^*(z)$. The Lyapunov exponent is given by

$$\lambda_n(A) = \frac{1}{pn + r} \log \left\{ |1 + g^{*'}(g^{*(-1)}(\xi_n))| \right\} \quad , \tag{2.4.13}$$

where $\xi_n = (A - A_n)/\alpha_n \gamma_n$. Thus we have

$$\tilde{\lambda}(z) = \frac{1}{p} \log \left\{ |1 + g^{*'}(g^{*(-1)}(z))| \right\} \tag{2.4.14}$$

as $n \rightarrow \infty$. Therefore, the existence of the fixed point function $g^*(z)$ means the existence of the fixed Lyapunov exponent $\tilde{\lambda}(z)$ and vice versa. Since $\alpha_n \propto n^{-2}$ and $\gamma_n \propto n^{-1}$, ξ_n is scaled by n^{-3}, which is consistent with the numerical result $\Delta A_n \propto n^{-3}$.

Up to now, we could not determine the form of g^* analytically. First we note that $g^*(0) = g^*(1) = 0$ and $g^*(z) \propto z^2$, $g^*(1-z) \propto (1-z)^2$ for $z \ll 1$. We expand $g^*(z)$ by Fourier series. Taking only the lowest order and noting the above property of $g^*(z)$, we have

$$g^*(z) = C(1 - \cos(2\pi z)) \quad . \tag{2.4.15}$$

We take this simple form, because it gives an analytic expression for $\lambda_n(A)$ and gives a qualitatively correct result. The numerical results (see Fig. 2.9) seem to support this simple approximation for the

(5n+4)-sequence at $D = 0.25$. Then we have $\sin^2 2\pi z^* = \xi_n(2C - \xi_n)/C^2$ and obtain $\lambda_n(A) = \frac{1}{(pn + r)} \log |1 - 2\pi\sqrt{\xi_n(2C - \xi_n)}|$. Thus, we get

$$\tilde{\lambda}(x) = \log\left\{ |1 - 4\pi C\sqrt{x(1 - x)}| \right\} . \qquad (2.4.16)$$

There are three possibilities for the behavior of $\lambda_n(A)$ according to the value of $\tilde{C} \equiv 2\pi C$ (curvature of the function $g^*(z)$).

Before going to the detailed discussion, we comment on the properties of Eq. (2.4.16):

a) $\lambda_n(A)$ is scaled by $\alpha_n\gamma_n \propto n^{-3}$,

b) $\tilde{\lambda}(x) \propto -x^{1/2}$ and $\tilde{\lambda}(1 - x) \propto -(1 - x)^{1/2}$ for $x \ll 1$,

c) $\tilde{\lambda}(x)$ is symmetric with respect to $x = 1/2$.

We note that the properties a) and b) are not dependent on the approximate choice (2.4.15) of $g^*(z)$. The property c) is valid if g^* is symmetric.

We discuss three possible cases in the next section.

§2.5 Classification of Period-Adding Sequences

As was discussed in §2.4, the similarity of period-adding sequence can be understood through $g^*(z)$. There are three possible cases according to the curvature of $g^*(z)$:

Case   I) Each (pn+r)-cycle does not have a superstable one and loses its stability through a tangent bifurcation.

Case II) Each (pn+r)-cycle has two superstable cycles and loses its stability through a tangent bifurcation.

Case III) Each (pn+r)-cycle loses its stability through period-doubling bifurcations. (That is, the Floquet multiplier changes from +1 to -1.)

If the original map is invertible, there is no superstable cycle and the only possible case is (I). Thus the period-adding sequence for the map (2.1.2) with $A < 1/(2\pi)$ corresponds to case (I).

We study these three cases using the function $g^*(z) = C(1 - \cos(2\pi z))$. (See Eq. (2.4.15).) Three cases are classified by the value $\tilde{C} \equiv 2\pi C$. We have plotted $\tilde{\lambda}(\xi)$ given by Eq. (2.4.16) for these three cases. (We use the notations $\tilde{\xi}_n = (A - A_n)/(2C\alpha_n\gamma_n)$.

(I) $0 < \tilde{C} < 1$

This condition corresponds to the case that the map is invertible. Thus, the attractor is torus or frequency locking. The Lyapunov exponent behaves like Fig. 2.10a). This exponent has a minimum at $\tilde{\xi}_n = 1/2$, and $\lambda_n^{min}(A) \propto 1/(pn + r) \propto 1/n$ for large n. The (5n+4)-sequence given in Fig. 2.6 belongs to this class.

We can fit the value C from the numerical result of $f_n(x)$. Using this value and Eq. (2.4.16) we can plot $\lambda_n(A)$, which is shown in Fig. 2.6, which agrees well with the numerical results. We note that the Schwarzian condition is broken both for the map with $A < 1/(2\pi)$ and $f^*(x)$ with $\tilde{C} < 1$.

(II) $1 < \tilde{C} < 2$

The (pn+r)-cycle has two superstable cycles at $\tilde{\xi}_n = (1 - \sqrt{1 - 1/\tilde{C}^2})/2$ and at $\tilde{\xi}_n = (1 + \sqrt{1 - 1/\tilde{C}^2})/2$. It loses its stability at $\tilde{\xi}_n = 1$ via a tangent bifurcation. The Lyapunov exponent behaves like Fig. 2.10b).

As an example of the case (II) we consider the map (2.1.2) with $D = 0.253$. At this parameter, the locking at 1/5 occurs at $A_{II} = 0.161625...$ $(> 1/2\pi)$. The sequence with $n/(5n + 4)$ appears for $A < A_{II}$. The Lyapunov exponent for this sequence behaves just like Fig. 2.10b). The approximate value of $\tilde{C}$ which we estimate from $f^{5n+4}(x)$ is 1.9. Thus this is an example for the case II). We note that there exists a chaotic region between (5n+4)-cycle and (5n+9)-cycle since A is larger than $1/2\pi$.

(III) $\tilde{C} > 2$

The (pn+r)-cycle loses its stability at $\tilde{\xi}_n = (1 - \sqrt{1 - 4/\tilde{C}^2})/2$ at which the Floquet multiplier crosses -1 and a period-doubling bifurcation occurs. At $\tilde{\xi}_n = (1 + \sqrt{1 - 4/\tilde{C}^2})/2$, the (pn+r)-cycle restores its stability via inverse period-doubling bifurcations. If $\tilde{C}$ is large

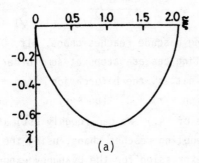

(a)

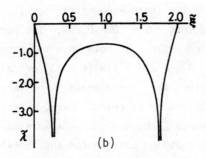

(b)

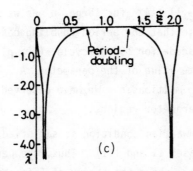

(c)

Fig. 2.10 The scaled Lyapunov exponent obtained from the phenomenological theory. The function

$$\log \left\{ \left| 1 - 2\tilde{C}\sqrt{\tilde{\xi}(1 - \tilde{\xi})} \right| \right\} \text{ (see Eq. 2.4.10)}$$

for (a) $\tilde{C} = 0.5$, (b) $\tilde{C} = 1.5$, (c) $\tilde{C} = 2.1$ are depicted.

enough and the interval $[(1 - \sqrt{1 - 4/\tilde{C}^2})/2,\ (1 + \sqrt{1 - 4/\tilde{C}^2})/2]$ is long, the period-doubling cascade reaches chaos. If $\tilde{C}$ is not large enough, the period-doubling cascade stops at some order and the period gets half by half.[*)] (That is, the bifurcation "(pn + r) → 2·(pn + r) → ... 2^{k-1}(pn + r) → 2^k(pn + r) → 2^{k-1}(pn + r) → ... (pn + r)" occurs as we increase the value of A.) We can roughly estimate the condition on whether the period-doubling reaches chaos, using the Feigenbaum's δ. (An example of the expression for the Lyapunov exponent is given in Fig. 2.10c).)

Let us consider some examples of the case III). As we increase the value of D from 0.254, the (5n+4)-sequence loses its stability through period-doubling bifurcations. For D = 0.254, the period-doubling bifurcation stops at some order. We have observed such sequences as 119 → 238 → 476 → 952 → 476 → 238 → 119 (n = 23) or 234 → 468 → 936 → 1872 → 936 → 468 → 234 (n = 46) for example. For larger values of D, period-doubling bifurcations go to chaos. Here, we note that the inverse cascade for (5n+4)-sequence ("chaos → ... → 2·(5n + 4) → (5n + 4)") can appear after the period-doubling cascade for the (5m+4)-cycle, where m is larger than n. At D = 0.256, for example, the sequence "(5n + 4) → 2·(5n + 4) → ... → chaos" appears before the sequence "chaos → ... → 2·(5 (n - 1) + 4) → 5(n - 1) + 4" for large n as we increase the value of A. We also note that the period-doubling cascade for (5m+4)-cycle and the inverse cascade for (5n+4)-cycle (m > n) can appear simultaneously at the same value of the parameter A. Thus the attractor is divided into multibasin. We have observed the multibasin phenomenon for various parameter regions.

We note that the Schwarzian condition is satisfied for the function (2.4.15) with $\tilde{C} > 1$ (Cases II and III). Thus, the period-doubling cascade in case III) always obeys the theory of Feigenbaum[2], if the doubling goes to infinity.

Thus, all the three cases appear for the map (2.1.2). We note that

---

*) The stop of period-doubling cascade was first pointed out by I. Tsuda (see Ref. 13)) when the Schwarzian condition is not satisfied (in our case it is satisfied).

this classification is independent of the special choice of $g^*(z)$ (Eq. (2.4.15)). For the sequence which appears at large A, the form (2.4.15) does not seem to be good, but the scaling and similarity hold just like in §2.3. We give a typical example in the next section.

§2.6 Period-Adding Sequence as Windows

For the cases II) and III), topological chaos already exists. Thus, the period-adding sequence in these cases can be regarded as a window sequence in the chaos. This type of the period-adding sequence appears in various system, which are not necessarily related to the transition from torus to chaos. In this section we study the period-adding sequence as windows.

First, we consider the period-adding sequence as A approaches $D (= 0.25)$, at which a stable fixed point appears via a tangent bifurcation. This sequence is given by

$$5 \to chaos \to 6 \to chaos \to 7 \to chaos \to \ldots \to 1$$
$$(p = s = 1, q = r = 0) \tag{2.6.1}$$

We have plotted in Fig. 2.11 the attractor for the map (2.1.2) with

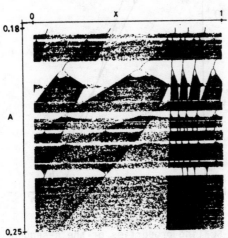

Fig. 2.11 The attractor for the map (2.1.2) with $D = 0.25$ as we change A from 0.18 to 0.25. We have plotted x_ns for $500 < n < 1300$ for each A.

42

D = 0.25 and for 0.18 < A < 0.25. Though the 5-period cycle changes
to chaos via intermittency, the n-cycle (n > 5) goes to chaos via period-
doubling bifurcations. Thus, the n-cycle sequence belongs to the case
of III) in 2.5.

First, we have checked $A_\infty - A_n \propto n^{-2}$ (see Fig. 2.12), which
agrees with our theory. Since the tangent bifurcation at A = 0.25 is
of the type "$x_{n+p} = x_n + ax_n^2 + \varepsilon$" (type I intermittency with z = 2),[1]
this is rather obvious. We have also checked $\Delta A_n \propto n^{-3}$ and $\Delta x_n \propto n^{-2}$.
We note that $\Delta x_n = A_\infty - A_n$, since $\varepsilon = A_\infty - A_n$ holds in this case.

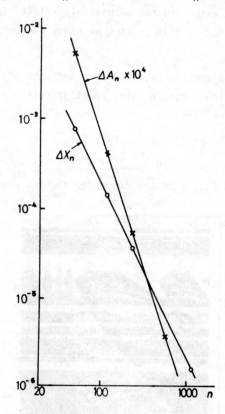

Fig. 2.12   Scaling properties of the n-sequence near
A ≤ D = 0.25. The quantities $\log (A_\infty - A_n)$
(which coincides with $\log \Delta x_n$ within the
numerical accuracy) and $\log \Delta A_n$ versus
$\log n$ are depicted.

The scaled Lyapunov exponent behaves like Fig. 2.10c) as is shown in Fig. 2.13, which seems to show the existence of the fixed Lyapunov exponent.

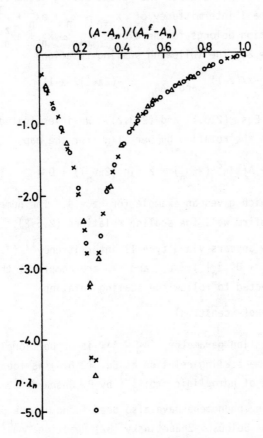

Fig. 2.13   Scaled Lyapunov exponent for the sequence with the rotation number $1/n$. The Lyapunov exponent is scaled by n^{-1} , while $(A - A_n)$ is scaled by n^{-3} . We have plotted the results for $n = 63(0)$, $118(X)$ and $554(\triangle)$.

The function $f^n(x) - x$, however, is very complicated even if we look only at the interval between the nearest periodic points. It has a lot of extremum points in the interval and the simple choice of the function (2.4.15) is impossible. This is due to the fact that A is large and $f^n(x)$ has a large number of extremum points.

So far, we have considered the period-adding sequence which corresponds to the type I intermittency of $x_{n+p} \simeq x_n + ax_n^2 + \epsilon$ [1]. If the tangent bifurcation belongs to the class $x_{n+p} \simeq x_n + ax_n^z + \epsilon$ instead of the case $z = 2$, the following scaling relations

$$A_\infty - A_n \propto n^{-z/(z-1)} \quad \text{and} \quad \Delta A_n \propto n^{-(2z-1)/(z-1)} \tag{2.6.2}$$

hold instead of Eqs. (2.3.1) and (2.3.2). We investigated a sequence of the windows with the rotation number $1/n$ for the map

$$\theta_{n+1} = \theta_n + A(\sin^2 (2\pi\theta_n) + 2 \sin (2\pi\theta_n)) + D \tag{2.6.3}$$

near $A = D$, which gives an example for $z = 4$. The numerical results for this map confirm well the scaling relations (2.6.2).

If a window appears via a type-II intermittency [1] (i.e., $z_{n+p} \simeq (1 + \lambda)z_n + O(|z_n|^2)$, z_n and λ are complex), the period-adding sequence is expected to follow the scaling relation

$$A_\infty - A_n \propto \exp(- \text{const. } n) \tag{2.6.4}$$

where the bifurcation parameter $(A_\infty - A)$ is proportional to Re λ. We note that the same scaling relation as Eq. (2.6.4) is found in a "period-adding" sequence of homoclinic orbits [3] by P. Gaspard. [4]

Period-adding phenomena have also been found in the periodic-chaotic transition of the Belousov-Zhabotinsky (BZ) reaction. [5]-[8] It will also be interesting to extend our formulation to area-preserving mappings [9] in connection with the devil's staircases of commensurate-incommensurate transitions. [10],[11]

§2.7 Scaling Properties at the Collapse of Tori - A brief review on a recent progress

Recently some important studies have appeared on the circle map (2.1.2). In this section these works are briefly reviewed on the emphasis on the critical phenomena at the collapse of tori.

Shenker[12] studied the sequence of lockings with the period of Fibonacci number for $|A| \leq 1/(2\pi)$ (see Fig. 2.2). He defined the value D_n by

$$f^{F_{n+1}}(\theta = 0.5; D = D_n) = F_n + 0.5 , \qquad (2.7.1)$$

for a given value of A. Here F_n is the Fibonacci sequence

0, 1, 1, 2, 3, 5, 8, 13, 21, 34, 55, 89, 144, 233, 377,
610, 987, 1597, 2584, 4181, 6765, 10946, 17711, 28657,
46368, 75025, 121393, ... $\qquad (2.7.2)$

which is constructed by $F_{n+1} = F_n + F_{n-1}$.[13] As n goes to infinity, the rotation number F_{n-1}/F_n approaches an irrational number $(\sqrt{5}-1)/2$, the inverse of the golden mean.

There is a historical reason to take this sequence of lockings. An irrational number ρ is approximated by a rational number using the following continued fraction expansion

$$\rho_k = 1/(n_1 + 1/(n_2 + 1/(n_3 ... + 1/n_k)...)) \qquad (2.7.3)$$

The "most irrational" number is considered to be as the one for which the approximation (2.7.3) is worst. The number is $\sigma_G^{-1} = 1/(1 + 1/(1 + 1/(1 + ... = (\sqrt{5} - 1)/2$ (or the one which has a tail of the continued fraction expansion $1/(1 + 1/(1 + ...).)$ The rational approximation for σ_G^{-1} is given by $\rho_k = F_k/F_{k+1}$, which is the reason to take the Fibonacci sequence. For the area-preserving systems, the KAM theory has already been established, which shows that the torus with the rotation number σ_G^{-1} is the last one to collapse as the perturbation is increased.[14] For a dissipative mapping, the torus with the rotation number σ_G^{-1} is not the last torus to collapse. The sequence with the rotation number F_k/F_{k+1}, however, has a very simple structure of continued fraction expansions and is useful to extract the characteristic feature of the collapse of tori.

The critical exponents are defined as follows;

$$D_\infty - D_n \propto \delta^{-n} \qquad (2.7.4)$$

$$f^{Fn}(\theta = 0.5, D = D_n) - (F_{n-1} + 0.5) \propto \alpha^{-n} . \qquad (2.7.5)$$

The second exponent represents the scaling for the distance between the nearest pair of the periodic points at $D = D_n$. Shenker found that δ and α take trivial values ($\delta = -\sigma_G^2$ and $\alpha = -\sigma_G$) for $A < 1/(2\pi)$, while they take nontrivial values ($\delta = -\sigma_G^Y$ and $\alpha = -\sigma_G^X$ where $Y \simeq 2.1644$ and $X \simeq 0.5269$) at $A = 1/(2\pi)$. The change of these exponents characterizes the collapse of tori.[12] Here the "trivial" values mean that they take the same values as for the linear case $A = 0$.

Feigenbaum, Kadanoff, Shenker[15] and Ostlund, Rand, Sethna, Siggia[16] independently constructed the renormalization group theory to explain the above critical phenomena. First, the function $\tilde{f}_n$ is defined by

$$\tilde{f}_n(\theta) = \alpha^n \left\{ f^{Fn+1}(\alpha^{-n}(\theta + 0.5); D = D_n) - (F_n + 0.5) \right\} . \qquad (2.7.6)$$

We assume that $\tilde{f}_n(\theta)$ approaches a fixed point function $f^*(\theta)$. Then the fixed point function satisfies the following set of functional equations;

$$f^*(x) = \alpha f^*(\alpha f^*(\alpha^{-2}x))$$

$$f^*(x) = \alpha^2 f^*(\alpha^{-1}f^*(\alpha^{-1}x)) . \qquad (2.7.7)$$

The equations (2.7.7) have the linear solution $f^*(x) = x - 1$, which gives the trivial critical exponents. The functional equations (2.7.7) and the trivial fixed point function were first obtained by I. Tsuda in connection with the BZ map.[17] At $|A| = 1/(2\pi)$, however, the linear solution is not relevant, because $f(\theta)$ has a cubic cusp point at $\theta = 0.5$, and $f^*(x)$ must be a function of x^3. The nontrivial fixed point function of this type was numerically obtained in the same manner as Feigenbaum's fixed point function for the period-doubling.[2] The relevant eigenvalue of the functional equation linearized around this fixed point function gives the exponent δ (α is simultaneously obtained with the fixed point function), which agrees well with the numerical results. (See Fig. 2.14 for the schematic representation of

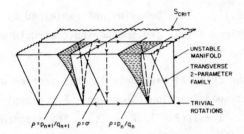

Fig. 2.14   Schematic representation of the global structure
of the dynamics of RG transformation in functional
space showing (a) both fixed points, (b) the
universal family corresponding to the unstable
manifold of the cubic fixed point, (c) a 2-parameter
family of analytic mappings of the circle which is
transverse to the stable manifold of the cubic
fixed point, and the relationship between the
regions ρ = const. for the universal and
transversal families. (cited from Ref. 16)

the renormalization group.)

If $|A|$ is slightly less than $1/2\pi$, the crossover phenomena occur,
that is the crossover from nontrivial critical phenomena caused by the
cubic fixed point to trivial critical phenomena by the linear fixed
point function. The quantities in (2.7.4) are represented as functions
of the single quantity εF_i^ν, where $\varepsilon = 1/2\pi - |A|$. The crossover
exponent ν agrees with 2x within the numerical error, which is
explained by the renormalization group theory.

The power spectrum at $A = 1/(2\pi)$ with the rotation number σ_G^{-1}
has a universal self-similar structure, which was found by S.J.
Shenker[12] and was explained by S. Ostlund et al.[16] The power spectrum
$|g(\omega)|^2$ is obtained from

$$g(\omega) = \lim_{L \to \infty} \frac{1}{L} \sum_{\ell=0}^{L-1} \exp(2\pi i \ell \omega)(f^\ell(0.5); \mod 1) . \qquad (2.7.8)$$

Near $\omega \sim 0$, $|g(\omega)|$ scales as ω, which shows

$$g(\omega) = -\sigma_G \, g(\omega/\sigma_G) + O(\omega^2) \quad \omega \to 0 \qquad (2.7.9)$$

(see Fig. 2.15). This behavior was explained by the above cubic fixed point function $f^*(x)$ and the periodic structure of the continued fraction for σ_G^{-1}.

All the critical exponents in this section depend on the character of the continued fraction expansion. For example, for the rotation number $\sqrt{2} - 1 = 1/(2 + 1/(2 + ...))$ the exponents take the values $x \approx 0.524$ and $y \approx 2.175$. Thus, it will be of importance to study the global character for the devil's staircase (see next section). The exponents agree if the rotation numbers have the same tail of the continued fraction expansion. Also, the exponents take the same values if the one-dimensional map $f(\theta) = \theta + D + A(\sin 2\pi\theta + 0.2 \sin 6\pi\theta)$ is chosen instead of (2.1.2).

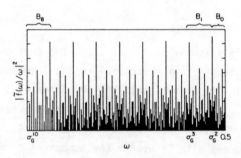

Fig. 2.15   Power spectrum of a time series for the map (2.1.2) with $A = 1/(2\pi)$ and $\rho = \sigma_G$ on a log-log plot. A normalization factor of ω^2 has been divided out of the power. The lines in each band B_j are in one to one correspondence and the associated complex amplitudes become universal for $j \to \infty$. Here $\tilde{f}(\omega) \equiv g(\omega)$. (cited from Ref. 16))

§2.8   Global Properties of the Devil's Staircase

In §2.7, critical phenomena at the collapse of tori were treated only for special rotation numbers. General global properties of the devil's staircase have recently been investigated. In the present section three works are briefly reviewed.

(I) Scalings of the devil's staircase[18]

Jensen, Bak, and Bohr[18] calculated the measures of the torus and lockings in the parameter space D for a given A $(\leq A_c = 1/(2\pi))$. First, the interval of D at which the locking to a rational rotation number p/q occurs is defined as $\Delta D(p/q)$ for a given A. M(A) = $\sum \Delta D(p/q)$ (summation is taking over all relatively prime integers p and q with q > p) is the measure for lockings. For A = 0, M(A) = 0. M(A) increases as A is increased, which is less than unity for A < A_c. If M(A) = 1, the devil's staircase is said to be "complete". Is the devil's staircase complete at A = A_c? (See Fig. 2.16a).)

To answer this question, Jensen et al. calculated the total width S(r) of all steps which are larger than r (i.e., the flat region larger than r in the staircase). Then N(r) = (1 - S(r))/r gives the number for the regions which are not flat in the devil's staircase. Numerical simulations show that (see Fig. 2.16b)),

$$N(r) \sim (1/r)^d \tag{2.8.1}$$

with $d = 0.87 \pm 3.7 \times 10^{-4}$. Thus, the space between steps of lockings vanishes as $1 - S(r) \sim r^{1-d}$ as $r \to 0$, which verifies the completeness of the staircase. The exponent d is the fractal dimension of the staircase (i.e., the dimension of the Cantor set of zero Lebesgue measure for the torus motion). The exponent seems to be universal if the original circle map has a cubic inflection point at the critical value of A (at which the map loses its invertibility).

The measure for torus (i.e., 1 - M(A)) in the parameter space D decreases as A approaches A_c. Numerical results show that

$$1 - M(A) \propto (1 - A/A_c)^\beta \tag{2.8.2}$$

with $\beta \sim 0.34 \pm 0.02$.

(II) Renormalization group for arbitrary rotation numbers[19][16]

The scaling analysis in §2.7 is restricted only to a rotation number which has a periodic tail of the continued fraction expansion.

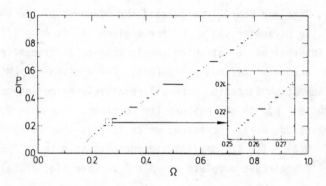

Fig. 2.16a) Rotation number ρ versus $\Omega \equiv D$ for the circle map at $A = A_c$. Steps with $\Delta(P/Q) > 0.0015$
are shown, while the inset shows intervals with
$\Delta > 0.00015$. (cited from Ref. 18))

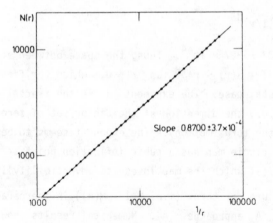

Fig. 2.16b) Plot of $\log_{10}N(r)$ versus $\log_{10}(1/r)$ for the
critical circle map. The slope of the
straight line yields $d = 0.8700 \pm 3.7 \times 10^{-4}$
(cited from Ref. 18))

What happens for a rotation with an arbitrary tail? Rand et al.[16] pointed out the possibility that the renormalization transformation has a strange attractor in that case. Farmer, Satija, and Umberger performed the renormalization procedure by Rand et al.[16] (see §2.7) numerically for the rotation number $\rho = 1/(n_1 + (1/n_2 + ...)...)$, with "random" n_i. Here "random" n_i is chosen by the procedure

i) pick an initial random seed ρ_0;

ii) iterate the Gauss map $\rho \to 1/\rho - [1/\rho]$ and

iii) $[1/\rho]$ for the i-th step gives n_i. (Here $[x]$ denotes the largest integer less than x).

The numerical result shows that the renormalization transformation has a strange attractor. The average values of the exponents α and δ (see §2.7) are roughly $\bar{\alpha} \sim 1.8 \pm 0.1$ and $\bar{\delta} \sim 15.5 \pm 0.5$.

(III) Renormalization of the Farey map and universal mode locking[20]

Quite recently Feigenbaum has constructed an interesting theory which relates lockings with period-doubling, under the influence of the idea by Cvitanovic and Shraiman.

The first step is to arrange possible p/q rotation numbers. As was shown in §2.2, all lockings are constructed by the so-called "Farey sum",

$$p/q \oplus p'/q' \equiv (p + p')/(q + q') \ . \tag{2.8.3}$$

We note that the Farey sum is constructed by a binary "Farey tree", (see Fig. 2.17a)).

Let us consider the rule for the Farey tree in the language of continued fraction expansions (see Fig. 2.17b)). With the notation $r \equiv [c_1, c_2, ... , c_n] = 1/\{c_1 + \{1/\{c_2 + ... 1/c_n\}...\}$, the rules are written as

$$0 : [c_1,...,c_{n-1},c_n+1] = [c_1,...,c_{n-1}] \oplus [c_1,...,c_{n-1},c_n]$$

$$1 : [c_1,...,c_{n-1},c_n-1,2] = [c_1,...,c_{n-1},c_n-1] \oplus [c_1,...,c_{n-1},c_n]$$

52

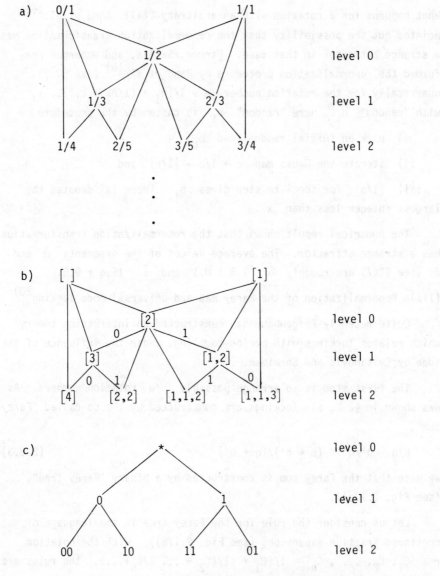

Fig. 2.17 a),b),c) Farey tree; (a) in the language of
rational number; (b) in the language
of continued fraction expansion;
(c) in the language of the binary rule.

Each k-th level gives rise to two daughters by the rule 0 or 1. Then, the Farey tree is written as in Fig. 2.17c) in the language of the rule (0, 1). (Each number $n_1...n_k$ shows the sequence of rules used to construct the rational number.)

The elements of the k-th level are labelled by r_0, r_1,...,r_{2^k-1} from left to right in ascending order. The next step is to construct a map on the unit interval which has these points as a periodic orbit of period 2^k. In order to do this, let us reorder the tree so that the reordered sequence at the k-th level is increasing denominators from left to right, while placing in nearby succession elements with similar continued fraction tails. This is done by changing the order of r_j and define $\tilde{r}_j$ according to the order of $n_1...n_k$ in Fig. 2.17c) (e.g., in the order of 00, 01, 10, 11 for k = 2 and $\tilde{r}_0$ = 1.4, $\tilde{r}_1$ = 3/4, $\tilde{r}_2$ = 2/5, and $\tilde{r}_3$ = 3/5). Then the one-dimensional map is obtained as

$$\tilde{r}_{m+1} = F(\tilde{r}_m) \ . \tag{2.8.4}$$

The third step is to calculate $F^{2^k}(x)$ and then to obtain the fixed point function $F^*(x)$ of the period-doubling using Feigenbaum's renormalization theory[2].

Up to this point, we have considered a linear circle map (pure rotation). For the nonlinear circle map (2.1.2) with $A < A_c$, there exists an analytic diffeomorphism $h(D)$ of the circle such that $h^{-1}_{(D)} \circ f_D \circ h(D) = R_\rho$ where f_D is the circle map and $R_\rho = \theta + \rho$ (simple rotation). Thus, the Farey map (2.8.4) can be written as a function of D (from a function of ρ) using the above $h(D)$. The map $\tilde{F}$ thus obtained has the same fixed point function F^* as for the map F.

The fixed point function F^* has all the information of scalings of the devil's staircase for $A < A_c$. See for details Ref. 20) (also for the extension to the critical case $A = A_c$).

§2.9 Supercritical Behavior of Disordered Orbits of a Circle Map

(i) In the present section we focus our attention on the super-critical behavior (namely for $A > A_c$) and study some properties of a disordered trajectory of a chaotic orbit (which will be defined in (iii)). Before going to the detailed study, we note the following properties of the map (2.1.2) $x_{n+1} = f(x_n) = x_n + A \sin(2\pi x_n) + D$ for $A > A_c$.

(1) As A is increased from A_c, chaos appears from the locking through usual scenarios (period-doubling[2] or intermittency[1] or crisis[24]). The measure (in the parameter space) of chaos increases for $A > A_c$ with the decrease of the measure of lockings from unity at $A = A_c$ as is shown in Fig. 2.18.

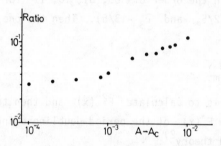

Fig. 2.18 The ratio of chaotic region in the parameter space D as a function of A. For each (A, D), we have made iterations of the map with the initial value $x_0 = 0.5$ and regarded the attractor as chaotic if $x_{n+100000} \neq x_{100000}$ for $1 < n < 50000$ within the error of 10^{-7}. For given A, D is changed from 0 to 0.5 by 2×10^{-3} (for $0.001 \leq A - A_c < 0.01$; 250 points of the D's are chosen) or by 10^{-3} (for $A - A_c < 0.001$; 500 points) and we have counted the number of the D's for which the attractor is chaotic. The ratio is defined by the number of the D's for chaos divided by 250 (for $0.001 \leq A - A_c < 0.01$) or 500 (for $A - A_c < 0.001$).

(2) The map (2.1.2) can have two stable attractors for $A > A_c$. The coexistence of two types of cycles or two types of chaos or a cycle and chaos is possible for several parameter regions. The

multibasin phenomenon is due to the existence of two critical points (i.e., the points where $f'(x) = 0$).

(3) The notion of "ordered" orbits was introduced by L.P. Kadanoff[25]. The orbits are in exactly the same order as the orbits for $x_{n+1} = x_n$ + (rotation number) and have a simple and beautiful property. The measure for such orbits, however, is zero and almost all orbits are disordered in the sense to be discussed in (iii). Thus, it will be of more importance to study disordered orbits in detail.

(4) There also exists a similarity for the lockings (or windows) for $A > A_c$. The two parameter self-similarity was shown by Glass et al.[21],[22] using the locus of superstable cycles and by Schell et al.[23].

In (ii), a phase diagram for the map with $A > A_c$ is given and the critical phenomena for $A \rightarrow A_c + 0$ are studied in a little detail. The notion of disordering is introduced in (iii) with the idea of induced map. In (iv), critical phenomena for the disordering property are investigated.

The crisis frequently appears in the circle map with $A > A_c$ and plays an important role in the change of the disordering property of the orbits in the circle map. In the appendix, the simplest case of the crisis, i.e., the map $x_{n+1} = 1 - ax_n^2$ with $a \rightarrow 2$ is studied using the induced maps. The period-adding sequence of superstable orbits is chosen to study the similarity of the orbits near the crisis.

(ii) Supercritical Similarity of the Circle Map

In Fig. 2.19, a part of the phase diagram of the map (2.12) for $A > A_c$ is given. Each basic cycle period-doubles by the increase of A. The strange figure of each locking (Arnold tongue) is understood from the viewpoint of the cusp bifurcation[23]. Here we consider the similarity and scaling among lockings in more detail. Each basic locking has a shape like Fig. 2.20 in the parameter space. The values δD and δA which are defined in Fig. 2.20, decrease as the period of

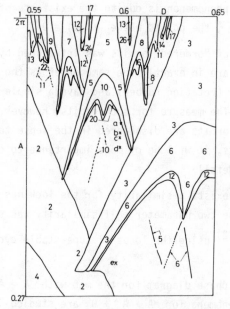

Fig. 2.19 Rough phase diagram for the circle map.
Numbers in the figure denote periods, while
chaos exists in the region without numbers.
Small structures, such as cycles with periods
larger than 26 are omitted. (Initial values
of iterations are $x_0 = 0.5$.)

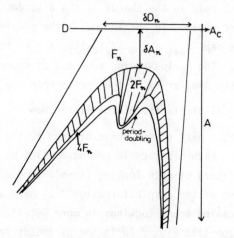

Fig. 2.20 Schematic representation of the Arnold
tongues for $A > A_c$.

the basic locking increases. Thus, the chaos appears immediately after A exceeds A_c for a locking with a long enough period. However, the measure of the locking with such a long period is small (i.e., δD decreases rapidly as the increase of the period). The measure of the chaos, therefore, grows up quite slowly, as is seen in Fig. 2.18.

In order to study the similarity, we consider the sequence of lockings with periods F_n[12),14)]. Here F_n is chosen to be the Fibonacci sequence (the rotation number is given by F_{n-1}/F_n) which is used as an approximation for the collapse of the golden mean torus. We note that δA is regarded as the index for the onset of chaos, since the chaos from a basic window appears by period-doubling at the parameter about $\delta A(1 + \delta_F^{-1} + \delta_F^{-2} + \ldots) \propto \delta A$, where δ_F is Feigenbaum's constant[2)].

δD_n and δA_n are defined as the values of δD and δA for the lockings with the rotation number F_{n-1}/F_n. As can be seen from Fig. 2.21 the relations

$$\delta A_n \propto F_n^{-a} \tag{2.9.1}$$

and

$$\delta D_n \propto F_n^{-d} \tag{2.9.2}$$

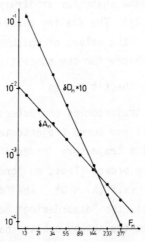

Fig. 2.21 Log δA_n and log δD_n vs log F_n.

are obtained with $a = 1.055(\pm 0.01)$ and $d = 2.165(\pm 0.01)$. (See Fig. 2.21.) The value a is close to the crossover exponent ν, which has been obtained by Shenker[12] and been explained by the RG theory[15],[16], as was given in §2.7.

The above result is explained as follows; as was shown in §2.7, the crossover exponent ν is defined by the postulate that the physical quantities are functions of the single quantity $(A - A_c)F_n^{\nu}$ for $A < A_c$. Thus the above result $\delta A \propto F_n^{-a}$ with $a = \nu$ shows that the crossover exponent takes the same values for $A > A_c$ and $A < A_c$. In generic cases, RG theories have a symmetry for super- and sub-critical regions, which brings about the same value for the exponents for both regions. The above result shows that the cubic fixed point function of the RG for the circle map at $A = A_c$ has the symmetry.

The exponent d takes the same value as y, where y is the exponent found by S.J. Shenker[12] as the convergence rate of the Fibonacci sequence at $A = A_c$ i.e., $\delta = -\sigma_G^y$ (see §2.7). The above agreement shows that the convergence rate is independent of the choice of the value of the Fibonacci-orbits (in Shenker's case the orbit which passes 0.5 is chosen[12] as is seen in §2.7).

To sum, the Arnold tongues have a similarity also for $A > A_c$, i.e., Fig. 2.20 takes a same shape for arbitrary F_n, if it is scaled by Eqs. (2.9.1) and (2.9.2). The scaling behavior is characterized by the exponents a and d, the values of which are consistent with the ones obtained by the RG theory for the subcritical region.

(iii) Disordering of the Chaotic Orbits

How are the chaotic trajectories characterized? The important difference between chaotic and torus trajectories lies in the "ordering" of trajectories. We call a trajectory "ordered"[25], if the nearby orbits do not change their order. Thus, an "ordered" trajectory cannot fall on the interval $I = \{x | f'(x) < 0\}$. The "disordering" means the loss of ordering. That is, the "disordering" occurs when an orbit falls on the interval I, where two nearby orbits change their order (see Fig. 2.22). The disordering of a trajectory in this sense is characterized by the ratio that the trajectory falls on the interval I.

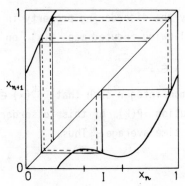

Fig. 2.22 An illustration of "disordering" of the orbits
for the circle map. Two nearby orbits (- and
---) change their order when they fall on the
interval I.

We define the disordering ratio d by the measure of the orbits
in I, i.e., by

$$d = \int_{x \in I} \rho(x) dx \qquad (2.9.3)$$

where $\rho(x)$ is the invariant measure.

First, we consider how the chaotic orbit with disordering appears.
When the period-doubling bifurcations from a p-cycle to chaos proceed,
the disordering ratio changes as

$$\frac{1}{p} = \frac{2}{(2p)} \rightarrow \underset{\underset{\text{doubling}}{\uparrow}}{\frac{1}{(2p)}} = \frac{2}{(4p)} \rightarrow \underset{\underset{\text{doubling}}{\uparrow}}{\frac{2a_{n-1}}{2^{n-1}p}} \rightarrow \underset{\underset{\text{doubling}}{\uparrow}}{\frac{a_n}{2^{n-1}p}} = \frac{2a_n}{2^n p} \cdots$$

where

$$a_n = (2^n - (-1)^n)/3 \; .$$

Chaos with the disordering ratio $d \sim 2/(3p)$ is born out of this
cascade, which appears in accordance with Feigenbaum's theory[2]. As
the nonlinearity A is increased, the disordering property begins to

differ from the case for a logistic map.

In order to study the disordering property in more detail, we introduce here the following induced map $F(x)$ on the interval I;

$$F(x) = f^k(x), \qquad x \in I \qquad\qquad (2.9.4)$$

where k is the minimum integer such that $f^k(x) \in I$. Furthermore, we introduce the distribution $P(k)$ of this "disordering time" k for one orbit with a long time average. Thus,

$$P(k) \propto \int_{x \in I_k} \rho(x)dx$$

holds, where $I_k \; (\subset I)$ is the region which satisfies $f^k(x) \in I$ and $f^m(x) \notin I$ for $m < k$.

The disordering ratio d is represented in terms of $P(k)$ as

$$d = \Sigma P(k)/k \; . \qquad\qquad (2.9.5)$$

Examples of the induced maps are given in Figs. 2.23a)-e), while the disordering time distributions (abbreviated as DTD hereafter) are illustrated in Figs. 2.24a)-d). The parameter values for these figures are shown in the phase diagram in Fig. 2.19.

Let us consider the change of disordering properties due to the increase of the nonlinearity A. After the period-doubling bifurcations form a p-cycle are completed, chaos with p-bands appears. As an example we consider the case $p = 5$ (with $D = 0.6$). Just after the accumulation of doubling cascades, the support of the invariant measure is restricted only in the region I_5 (i.e., in the region $f^5(x) \in I$ and $f^m(x) \notin I$ for $m < 5$). See Fig. 2.23a) for the induced map. Thus the DTD $P(k)$ is nonzero only for $k = 5$.

As the nonlinearity A is increased, the support of the measure increases into the regions I_5 and I_{10} and $P(k) \neq 0$ for $k = 5$ and 10 (see Figs. 2.23b) and 2.24a)). As the nonlinearity A is increased further, the disordering time $5 \times n$ $(n = 1,2,3,4,...)$ appears successively (see Figs. 2.23c) and 2.24b) for example) and at some critical point all the disordering times of the form $5 \times n$ exist, where the "crisis" of the chaos with 5-bands occurs. The above mechanism of the evolution is independent of p (period of the band). The essential mechanism of the evolution of DTD near the crisis is understood by the logistic map $x_{n+1} = ax_n(1 - x_n)$ at $a \sim 2$, which is shown in the Appendix.

After the crisis of a 5-chaos occurs, the orbit can go out of the region $\overset{\infty}{\underset{n=1}{U}} I_{5n}$. When the parameter A is increased and exceeds the region of the doubling cascade of $m \times 2^n$ in the phase diagram, a new disordering time m (and $k \times m$ (k = 1,2,3,...) sucessively) appears. For example, at the point 'd' (see Fig. 2.19), the disordering time 8 appears (see Figs. 2.23d) and 2.24c)). As A is increased further, the disordering times $8n + 3k$ appear successively, till the disordering time 3 appears (see Figs. 2.23e) and 2.24d)). The disordering property evolves in this way as the increase of the nonlinearity.

(iv) Supercritical Behavior of Disordering Property

Here, we consider DTD and induced maps in more detail, especially focusing on the properties at $A \rightarrow A_c$.

As is shown in the Appendix, the DTD behaves as

$$P(nk) \propto r^{-n} \tag{2.9.6}$$

when the crisis of a k-band chaos occurs. Here, the value r is given by the instability exponent

$$\left| f^{k\,\prime}(x_i) \right| = \left| \overset{k}{\underset{i=1}{\Pi}} f'(x_i) \right| , \tag{2.9.7}$$

where x_i is an unstable periodic point (with period k).

Let us take into account the similarity of windows in (ii) and consider the case when each F_k-band (F_k is the Fibonacci sequence) chaos shows a crisis. The parameters A and D are taken so that the similarity holds. Then at each value for the chaos with bands with Fibonacci numbers

$$P(nF_k) \sim P(F_k) \times r^{-n} \tag{2.9.8}$$

62

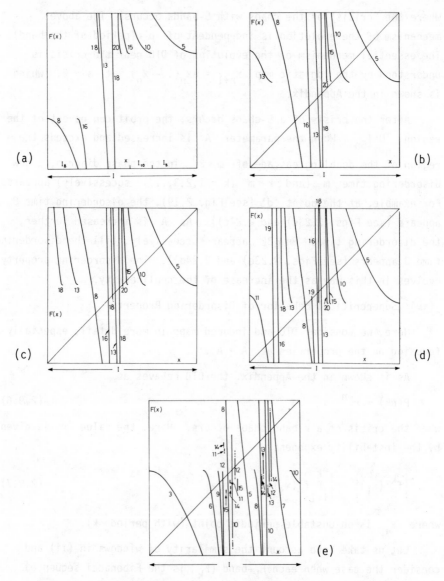

Fig. 2.23 Induced map F(x) (see Eq. (2.9.4)) with the
disordering time k, where the time k is
shown up to 20 for (a)~(d) and up to 15 for
(e). D is chosen to be 0.6.

(a) A = 0.196,   (b) A = 0.204,
(c) A = 0.2058, (d) A = 0.210,
(e) A = 0.254.

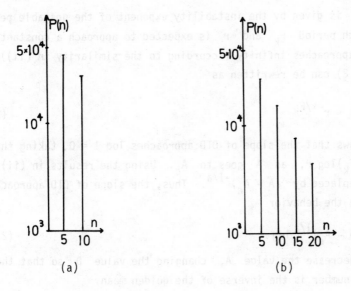

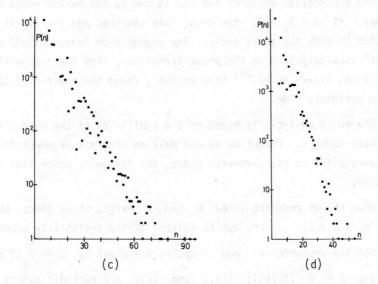

Fig. 2.24 Distribution of disordering times P(n),
which is obtained from 50000 iterations
of the circle map with D = 0.6. (Initial 10000 iterations are dropped.)

(a) A = 0.204,   (b) A = 0.2058,
(c) A = 0.210,   (d) A = 0.254.

where r is given by the instability exponent of the unstable periodic point with period F_k and r is expected to approach a constant value as F_k approaches infinity (according to the similarity in (ii)). The Eq. (2.9.8) can be rewritten as

$$P(x) \sim r^{-x/F_k} \qquad (2.9.9)$$

which shows that the slope of DTD approaches $\log 1 = 0$, taking the form of $-(1/F_k)\log r$, as A goes to A_c. Using the results in (ii), F_k may be replaced by $(A - A_c)^{-1/a}$. Thus, the slope of DTD approaches 0 by taking the behavior

$$- (A - A_c)^{1/a} \log r \qquad (2.9.10)$$

when we decrease the value A, changing the value D so that the rotation number is the inverse of the golden mean.

The exponential decay of the DTD is due to the Markov property of the map. At $A = A_c + 0$, the decay rate vanishes and the DTD is expected to show the power decay. The change from "exponential" to "power" is widely seen in the phase transition. For the intermittent transition, Aizawa et al.[26] have recently found the power distribution of the residence time.

The above analysis is based on the similarity of the band with Fibonacci numbers. Though we cannot make an accurate argument for an arbitrary point in the parameter space, the following properties should be noted:

(1) When the disordering times k and ℓ exist, other times $mk + n\ell$ $(m, n = 1, 2, \ldots)$ are apt to exist. If the instability exponents for the unstable k- and ℓ-cycles are given by $\alpha = \prod_{i=1}^{k} |f'(x_i)|$ and $\beta = \prod_{j=1}^{\ell} |f'(x_j')|$ ($\{x_i\}$ and $\{x_j'\}$ are periodic points for k- and ℓ-cycles respectively), the DTD is given approximately by

$$P(mk + n\ell) \sim \alpha^{-m}\beta^{-n} \qquad (2.9.11)$$

(2) For an arbitrary cycle with the period k, $P(k)$ is roughly proportional to $r_k^{-1} = |\prod_{i=1}^{k} f'(x_i)|^{-1}$ ($\{x_i\}$ are periodic points)[27]. Roughly speaking, $r_k \sim <f'>^k$, where $<f'>$ is some average of $f'(x)$. Since $<f'>$ is expected to grow as A is increased, the decay rate r gets larger with the increase of A, as is typically seen in Figs. 2.24c) and 2.24d) and 2.25. (Note the change of the scales between Figs. 2.25a) and b). The change of slope between the two figures caused by the reduction of $A - A_c$ is consistent with the scaling relation $(A - A_c)^{1/\nu}$.) As A approaches A_c, $<|f'|>$ is expected to approach 1. Thus, DTD for the chaos near $A = A_c$ (with arbitrary rotation number) is expected to show the power decay.

Of course, the decay of DTD as a function of time is not monotonic. It has many peaks and is not simple at all (see Figs. 2.24 and 2.25). We also note that smaller disordering times are successively inhibited as A approaches A_c (see Figs. 2.25a) and b)).

(3) It will be important to construct a symbolic dynamics $I_j \rightarrow I_k$. From the transition probability between the states I_j and I_k, we can understand the disordered property of chaotic trajectories[28].

As another example of supercritical behavior, let us consider the increase of the Lyapunov exponent as a function of $(A - A_c)$. In Fig. 2.26, maximum Lyapunov exponent for $0.606 < D < 0.607$ is plotted as a function of $(A - A_c)$, where the rotation number of the map is close to $(\sqrt{5} - 1)/2$. Numerically, the max L was obtained as the maximum of $L(0.606 + 2i \times 10^{-5})$ for $i = 1,2,...,50$. If the small dips are neglected, the increase of max L is roughly proportional to $(A - A_c)^{1/\nu}$ which again justifies the above argument on the supercritical behavior.

In connection with the supercritical behavior of a circle map, the following point should be noted:

The exponents for the subcritical regions depend on the structure of continued fraction expansions of the irrational rotation number.

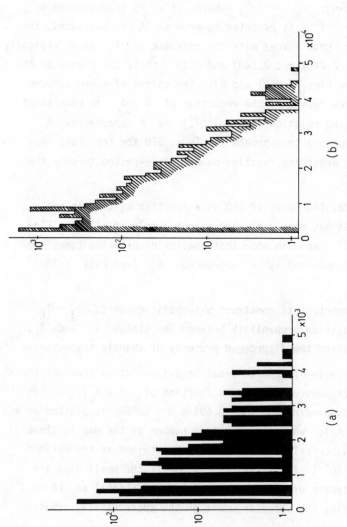

Fig. 2.25 Histogram of the distribution of disordering times $P(n)$, which is obtained from 10^7 iterations of the circle map. (Initial 10^4 iterations are dropped.)

(a) $A = A_c + 0.0008$ and $D = 0.6065$: Longitudinal axis is the summation of $P(n)$ for the interval $(n \times 10^2, (n+1) \times 10^2)$.

(b) $A = A_c + 0.0001$ and $D = 0.6066$: Longitudinal axis is the summation of $P(n)$ for the interval $(n \times 10^3, (n+1) \times 10^3)$.

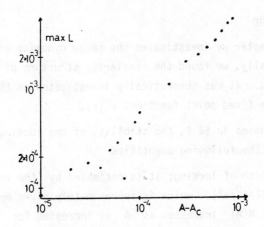

Fig. 2.26 Maximum of the Lyapunov exponent L(D) for 0.606 < D < 0.607 as a function of A. Lyapunov exponent was calculated from the data x_n's of the map (2.1.2) for

$$10^4 < n < 6 \times 10^4,$$ with the initial value $x_0 = 0.5.$

Thus, it is expected that the speed of the collapse of tori (it may be characterized by the exponent a in (ii) where $\delta A_n \propto F_n^{-1/a}$) depends on the irrationality of the torus. According to the results by S.J. Shenker[12], a = 1.0476... for the torus with the rotation number 1/(2 + (1/2 + (1/2 + ...), since a = ν is strongly suggested in (ii) (note that the exponent is smaller than the one for the golden mean torus). Thus, the golden mean torus collapses faster than the torus with 1/(2 + (1/2 + (1/2 + ...) in the above sense, which is quite con- trary to the well-established result for the area-preserving mappings[14], where the golden mean torus is the last KAM to collapse. It will be of importance to check the conjecture that the golden mean torus is the first to collapse in the dissipative mappings (i.e., a = ν takes its maximum for the rotation number with the tail of 1/(1 + /(1 + /(1 + ...))). In some sense, the conjecture is rather natural, since the locking with a smaller period collapses at a larger value of A (by period-doublings) in the circle map, while the stochasticity around the resonance with a smaller period plays a more essential role for the collapse of tori in area-preserving systems.

§2.10 Discussion

In this chapter we investigated the phase dynamics of the torus motion. Especially, we found the similarity structure of the period-adding sequences. It was theoretically investigated on the basis of the existence of the fixed point function $f^*(x)$.

As is discussed in §2.4, the stability of the locking to p-cycle is estimated by the following quantities

(i) the width of locking; it is estimated by (the variation of $f^p(\theta)/(\partial f^p/\partial A)$, which is approximately given by $N(A) \cdot p^{-3}$, where $N(A)$ increases as A is increased for $A < 1/2\pi$.

(ii) Lyapunov exponent, which is approximately proportional to p^{-1}.

From this point of view, the locking with a shorter period is easier to be observed. Thus, the locking with $(q + s)/(p + r)$ is most feasible to be observed between the lockings with q/p and s/r, and the locking with $(2q + s)/(2p + r)$ is most feasible to be observed between the lockings with q/p and $(q + s)/(p + r)$, and so on. This is the reason why the period-adding sequence (i.e., the sequence with $(nq + s)/(np + r)$) is feasible to be observed.

The period-adding sequence is observed in various systems which have no direct connection with the transition from torus to chaos. For example, the period-adding sequence of windows $2 \to 3 \to 4 \to \ldots \to 1$ is observed (up to period 5) for the BZ reaction with flow.[5)-8)] Since the chaos and the bifurcation structure are well explained by a one-dimensional mapping,[7),8)] the analysis based on the tangent bifurcation will be possible to give the scaling and similarity structure of the above sequence. The period-adding sequence also appears in the sequence of homoclinic orbits,[4)] and in the sequence of windows near the crisis and in the windows among the chaos of the Duffing's equation.[29)] Thus, the period-adding sequence with similarity structure is a rather general feature in the bifurcation of nonlinear systems.

Recently devil's staircases have been investigated in a lot of systems. The commensurate-incommensurate transition in solid state physics is a typical example. S. Aubry[10] considered this problem using the Frenkel-Kontoreva model

$$r_{n+1} = r_n - \frac{k}{2\pi} \sin (2\pi\theta_n)$$

$$\theta_{n+1} = \theta_n + r_{n+1} ,$$

(2.10.1)

which has been investigated as a "standard mapping"[30] in area-preserving systems. There are elliptic and hyperbolic periodic points. Around the elliptic points, there are KAM tori. We can investigate the period-adding sequences of elliptic or hyperbolic cycles, to study the similarity of scaling relations.

Spin systems with anisotropic next nearest neighbor coupling can show the devil's staircase, which have been extensively studied by P. Bak et al.[11] and M.E. Fisher et al.[31] If the lattice structure is a Cayley tree or the hierarchical lattice, the Bethe approximation or the Migdal-Kadanoff renormalization group is exact. In these cases, dissipative mappings are made used of to calculate the thermodynamic quantities.[32]

The behavior of a one-dimensional Schrödinger equation in a quasi-periodic potential is also an interesting problem in solid-state physics. This problem has been studied by the following two-dimensional mapping

$$a_{n+1} + a_{n-1} = Ea_n + V(n)a_n ,$$

(2.10.2)

where tight binding model is used and $V(n)$ is a periodic function with an irrational period (e.g., $V(n) = V_0 \cos (2\pi\sigma n)$, σ = irrational). When the strength of quasiperiodic potential V reaches a critical value, the energy spectrum of the discrete Schrödinger Eq. (2.10.2) shows a Cantor structure. This problem was studied by Hofstadter,[33] S. Aubry,[34] and Kohmoto[35],[36] and others. It has recently been noted[36],[37] that the critical phenomena at this transition have a quite similar structure to the transition treated in §2.7.

The noise effect on the transition from torus to chaos with lockings is also an important problem to study. If the noise is added, the small structures of lockings cannot be observed. It will be of interest to study the scaling property between the strength of noise and the stability of the locking.[38] The Lyapunov exponent for the chaotic state near $A \geq 1/(2\pi)$ decreases when the noise is added. It is considered to be due to the effects of neighboring windows. Thus, this phenomenon will not be related to a noise-induced order by Matsumoto and Tsuda.[39]

After the similarity of the period-adding sequence is described in §§2.3-2.6, the renormalization group approach at the collapse of tori is reviewed in §2.7. It is an interesting framework which reveals the mechanism of the collapse of tori, though the detailed observation of this theory remains rather difficult. §2.8 treats some further studies on the scalings of devil's staircase.

In §2.9, we attempted to characterize the property of chaos for the circle map (2.1.2). We introduced the notion "disordering", which shows how the orbit loses its ordering. The detailed study of the property of chaos, however, is left to the future.

In this chapter, we restricted ourselves only to the phase motion of torus and considered the one-dimensional mapping. Our results in this chapter, however, hold also for higher-dimensional systems, if we consider only the phase part of the torus. We studied, for example, the following two-dimensional mappings

$$x_{n+1} = Ax_n + (1 - A)(1 - Dy_n^2), \quad y_{n+1} = x_n \tag{2.10.3}$$

and

$$x_{n+1} = 1 - Ax_n^2 + D(y_n - x_n), \quad y_{n+1} = 1 - Ay_n^2 + D(x_n - y_n) . \tag{2.10.4}$$

The structure of lockings and the similarity structure are well explained by the one-dimensional mapping (2.1.2), though a new phenomenon "locking with symmetry breaking" occurs for the model (2.10.4), which will be described in the next chapter. For a flow system (i.e., the system governed by differential equations), locking structure with similarity is also understood by the one-dimensional mapping (2.1.2). The period-adding sequence of lockings was investigated by M. Sano and Y. Sawada for a coupled Brusselator model,[40] while R.V. Buskirk and C. Jeffries found the sequence in the experiments for a coupled p - n junction system.[41]

Before closing this chapter, we make a sketch of recent experiments on the transition from torus to chaos. The experiments on this topic have mainly been performed by i) Rayleigh-Bénard convection, ii)Taylor-Couette vortices and iii) Josephson junction. The Rayleigh-Bénard convection has been most extensively investigated. An excellent summary about the routes to turbulent convection was given by J.P. Gollub and S.V. Benson,[42] where the transition "torus → locking → chaos" was also shown. The velocity records and the power spectra are shown in Fig. 2.27 where the transition is remarkably seen. The locking at $f_2/f_1 = 7/3$ is shown in Fig. 2.28. The power spectra have also been measured by a lot of physicists, including G. Ahlers and R.P. Behringer,[43] A. Libchaber and his group,[44] M. Sano and Y. Sawada[45] and others. These results show some lockings at simple rational frequency ratios, though the detailed experiments which confirm the scaling relations and critical phenomena are left to the future. The noisy spectra increase as the Rayleigh number is increased above the value of the collapse of tori (via a locking) (see Fig. 2.29 and also Fig. 2.30), though the theoretical analysis remains incomplete for the supercritical region and cannot predict the behavior of noise power.

Recently Poincaré map of the flow has been obtained by Bergé,[46] M. Sano and Y. Sawada[45] using the method Takens. From the Poincaré map, a one-dimensional mapping for the phase motion has also been obtained. Examples of these mappings are shown in Fig. 2.31. These results seem to confirm the validity of the analysis based on the circle map (2.1.2).

The Taylor-Couette flow also gives an example of the transition from torus to chaos accompanied by lockings.[47] Josephson junctions are also good candidates to investigate the transition from torus to chaos. The equation which governs the system is given by

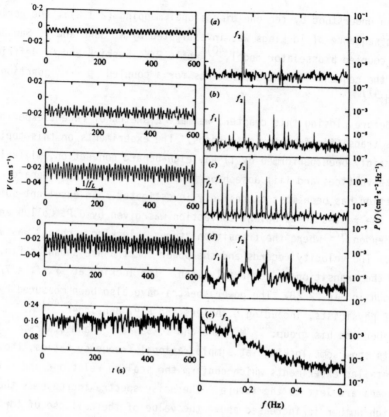

Fig. 2.27 Velocity records and power spectra showing the
sequence of instabilities leading to non-
periodic flow (route Ia). The sequence consists
of: (a) a periodic state with a single peak
and its harmonics, R/R_C = 31.0; (b) a quasiperiodic state with two incommensurate frequencies f_1 and f_2 and many of their linear combinations, R/R_C = 35.0; (c) phase locking at the
integer ratio f_2/f_1 = 9/4, R/R_C = 45.2; (d) a non-
periodic state with relatively sharp peaks just
above the onset of noise, R/R_C = 46.8; and (e) a
strongly non-periodic state with no sharp peaks
showing the broadband noise far above its onset,
R/R_C = 65.4.

(cited from Ref. 42; aspect ratio = 3.5, Prandtl
number = 5.0)

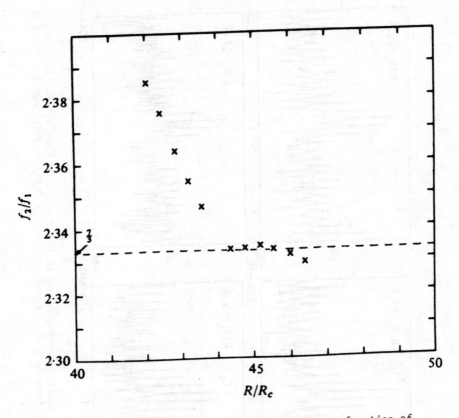

Fig. 2.28 Frequency ratio f_2/f_1 plotted as a function of R for route Ia. The ratio has a step indicating phase locking in the range $44.4 < R/R_c < 46.0$.

The errors in the values are approximately equal to the size of the symbols. (cited from Ref. 42; aspect ratio = 3.5, Prandtl number = 5.0)

74

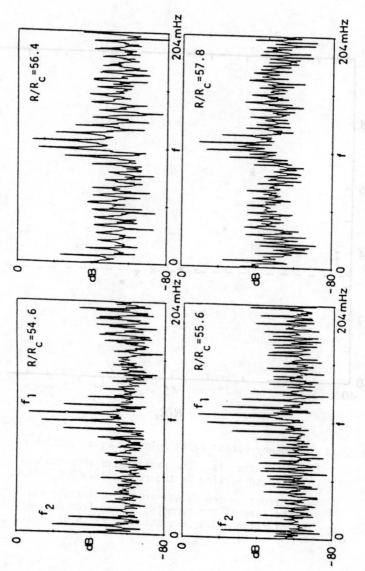

Fig. 2.29 Power spectra corresponding to the transition from 2-torus to chaos.
(cited from Ref. 45); aspect ratio $\Gamma = 3.0$, Prandtl number $P = 6.0$

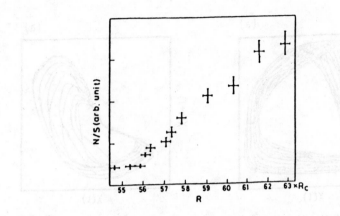

Fig. 2.30   Integrated noise power versus Rayleigh number.
(cited from Ref. 45); aspect ratio $\Gamma = 3.0$,
Prandtl number $P = 6.0$)

$$\ddot{\theta} + \gamma g(\theta)\dot{\theta} = a \sin (2\pi\theta) + I_{dc} + I_{ac} \cos \omega t \ . \qquad (2.10.5)$$

Simulations of the equation (2.10.5) have been performed by B.A.
Huberman[48] et al., N.F. Pedersen et al.,[49] M.H. Jensen[50] and others,
which show the transition from torus to chaos according to the phase
dynamics. The devil's staircase of the lockings has also been partly
observed in the experiment.[51]

The rhythm of autonomous biological oscillators is affected by
periodic perturbation. If the perturbation is strong enough, frequency
lockings and chaos appear. Onset of chaos via lockings has recently
been observed experimentally for stimulated cardiac cells[52] and
neurons.[53],[54]

To sum, the transition from torus to chaos accompanied by lockings
has been observed in various experiments, which are well explained by
the one-dimensional map for the phase motion. The quantitative compar-
ison between theories and experiments, however, is left to the future.

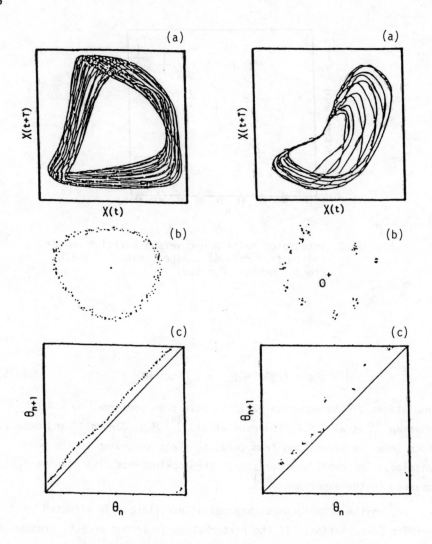

Fig. 2.31 (i) Quasi-periodic state, (ii) Phase-locked state.
(a) Two-dimensional projection of the orbit;
(b) Poincaré section; (c) One-dimensional return
map ($\Gamma = 3.0$). (cited from Ref. 45); aspect
ratio $\Gamma = 3.0$, Prandtl number $P = 6.0$

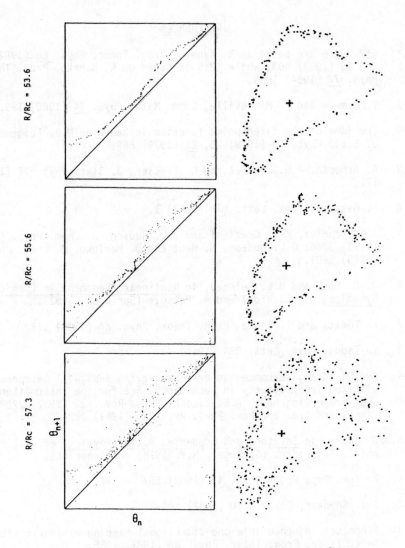

Fig. 2.31 (iii) Poincaré section and one-dimensional return
map of torus in the vicinity of transition
point to chaos (Γ = 2.95). (cited from
Ref. 45); aspect ratio Γ = 2.95, Prandtl
number P = 6.0)

REFERENCES

*) §§2.3-2.6 are based on K. Kaneko, Prog. Theor. Phys. <u>68</u> (1982) 669 and <u>69</u> (1983) 403, while §2.9 is based on K. Kaneko, Prog. Theor. Phys. <u>72</u> (1984) 1089.

1. Y. Pomeau and P. Manneville, Comm. Math. Phys. <u>74</u> (1980) 189.

2. The idea of the fixed point function is based on M.J. Feigenbaum, J. Stat. Phys. <u>19</u> (1978) 25, <u>21</u> (1979) 669.

3. A. Arneodo, P.H. Coullet and C. Tresser, J. Stat. Phys. <u>27</u> (1982) 171.

4. P. Gaspard, Phys. Lett. <u>97A</u> (1983) 1.

5. R.A. Schmitz, K.R. Graziani and J.L. Hudson, J. Chem. Phys. <u>67</u> (1977) 3040; J.L. Hudson, M. Haut and D. Marinko, J. Chem. Phys. <u>71</u> (1979) 1601.

6. J.-C. Roux and H.L. Swinney, in <u>Nonlinear Phenomena in Chemical Dynamics</u>, ed. C. Vidal and A. Pacault (Springer, 1981).

7. K. Tomita and I. Tsuda, Prog. Theor. Phys. <u>64</u> (1980) 1138.

8. I. Tsuda, Phys. Lett. <u>85A</u> (1981) 4.

9. Period-adding sequences in area-preserving maps will be connected with the intermittency in such maps. See for the intermittency in area-preserving maps, A.B. Zisook, Phys. Rev. <u>A25</u> (1982) 2289, A.B. Zisook and S.J. Shenker, Phys. Rev. <u>A25</u> (1982) 2824.

10. S. Aubry, in <u>Solitons and Condensed Matter Physics</u>, ed. A.R. Bishop and T. Schneider, (Springer, N.Y. 1978) and preprints.

11. P. Bak, Rep. Prog. Phys. <u>45</u> (1982) 587.

12. S.J. Shenker, Physica <u>5D</u> (1982) 405.

13. Fibonacci sequence in a one-dimensional mapping was first studied by I. Tsuda, Prog. Theor. Phys. <u>66</u> (1981) 1985.

14. J.M. Greene, J. Math. Phys. <u>9</u> (1968) 760, <u>20</u> (1979) 1183; L.P. Kadanoff, Phys. Rev. Lett. <u>47</u> (1981) 1641.

15. M.J. Feigenbaum, L.P. Kadanoff and S.J. Shenker, Physica <u>5D</u> (1982) 370.

16. S. Ostlund, D. Rand, J. Sethna and E.D. Siggia, Physica <u>8D</u> (1983) 303.

17. See Ref. 13).

18. M.H. Jensen, P. Bak and T. Bohr, Phys. Rev. Lett. $\underline{50}$ (1983) 1637; Phys. Rev. $\underline{A30}$ (1984) 1960.

19. J.D. Farmer and I.I. Satija, Phys. Rev. $\underline{A31}$ (1985) 3520; J.D. Farmer, I.I. Satija and D. Umberger, to be published.

20. M.J. Feigenbaum, to be published; Talk at the Workshop "Dynamic Days" (1985, La Jolla).

21. L. Glass and R. Perez, Phys. Rev. Lett. $\underline{48}$ (1982) 1772.

22. R. Perez and L. Glass, Phys. Lett. $\underline{90A}$ (1982) 441.

23. M. Schell, S. Fraser and R. Kapral, Phys. Rev. $\underline{A28}$ (1983) 1637.

24. C. Grebogi, E. Ott and J. Yorke, Phys. Rev. Lett. $\underline{48}$ (1982) 1507.

25. L.P. Kadanoff, J. Stat. Phys. $\underline{31}$ (1983) 1.

26. Y. Aizawa, Prog. Theor. Phys. $\underline{72}$ (1984) 659; Y. Aizawa and T. Kohyama, in Chaos and Statistical Methods, (Springer, 1984, ed. Y. Kuramoto).

27. T. Kai and K. Tomita, Prog. Theor. Phys. $\underline{64}$ (1980) 1532.

28. J. Guckenheimer, Physica $\underline{1D}$ (1980) 227. Y. Aizawa, Prog. Theor. Phys. $\underline{70}$ (1983) 1249.

29. S. Sato, M. Sano and Y. Sawada, Phys. Rev. $\underline{A28}$ (1983) 1654.

30. B.V. Chirikov, Phys. Rep. $\underline{52}$ (1979) 263.

31. W. Selke and M.E. Fisher, Phys. Rev. $\underline{B20}$ (1979) 257; M.E. Fisher and W. Selke, Phys. Rev. Lett. $\underline{44}$ (1980) 1502.

32. S.R. MacKay, A.N. Berker and S. Kirkpatrick, Phys. Rev. Lett. $\underline{48}$ (1982) 767; S. Inawashiro, C. Thompson and G. Honda, preprint.

33. D.R. Hofstadter, Phys. Rev. $\underline{14}$ (1976) 2239, see also P.G. Harper, Proc. Phys. Soc. $\underline{A68}$ (1955) 874.

34. S. Aubry and G. André, Ann. Israel Phys. Soc. $\underline{3}$ (1980) 133.

35. M. Kohmoto, Phys. Rev. Lett. $\underline{51}$ (1983) 1198.

36. M. Kohmoto, L.P. Kadanoff and C. Tang, Phys. Rev. Lett. $\underline{50}$ (1983) 1870.

37. S. Ostlund, R. Pandit, D. Rand, H.J. Schellnhuber and E.D. Siggia, Phys. Rev. Lett. $\underline{50}$ (1983) 1873.

38. M.J. Feigenbaum and B. Hasslacher, Phys. Rev. Lett. $\underline{49}$ (1982) 605.

80

39. K. Matsumoto and I. Tsuda, J. Stat. Phys. _31_ (1983) 87.

40. M. Sano and Y. Sawada, Phys. Lett. _97A_ (1983) 73.

41. R.V. Buskirk and C. Jeffries, Phys. Rev. _A31_ (1985) 3332.

42. J.P. Gollub and S.V. Benson, J. Fluid Mech. _100_ (1980) 449.

43. G. Ahlers and R.P. Behringer, Suppl. Prog. Theor. Phys. _64_ (1978) 186.

44. J. Maurer and A. Libchaber, J. de Physique Lett. _40_ (1979) 419, A. Libchaber, S. Fauve and C. Laroche, Physica _7D_ (1983) 73.

45. M. Sano and Y. Sawada, to appear in Chaos and Statistical Methods (ed. Y. Kuramoto, Springer 1984).

46. P. Bergé, Physica Scripta _T1_, (1982) 71.

47. H.L. Swinney, Suppl. Prog. Theor. Phys. _64_ (1978) 164.

48. B.A. Huberman, J.P. Crutchfield, and N.H. Packard, Appl. Phys. Lett. _37_ (1980) 751.

49. N.F. Pedersen and A. Davidson, Appl. Phys. Lett. _39_ (1981) 830.

50. M.H. Jensen et al., preprint.

51. R.F. Miracky, J. Clarke, and R.H. Koch, Phys. Rev. Lett. _50_ (1983) 856.

52. M.R. Guevara, L. Glass, A. Shrier, Science _214_ (1981) 1350.

53. H. Hayashi, S. Ishizuka and K. Hirakawa, J. Phys. Soc. Jpn _52_ (1983) 344.

54. G. Matsumoto et al., J. Theor. Neurobiology _3_ (1984) 1.

Appendix

In the present appendix, we consider the change of the chaotic orbits near the crisis,[24] using the induced map and the period-adding sequence. The logistic map

$$x_{n+1} = F(x_n; a) = 1 - ax_n^2 \qquad (A.1)$$

near $a = 2$ is taken as the simplest and essential case. We construct the induced map for (A.1) by choosing the interval $(0, F(0))$. In Figs. 2.32a)-c), some examples of the induced maps are shown. At $a = a_n$, a new number "n" appears (it corresponds to the disordering time in §2.9. The orbits for $a > a_n$ and for $a < a_n$ differ in the following sense.

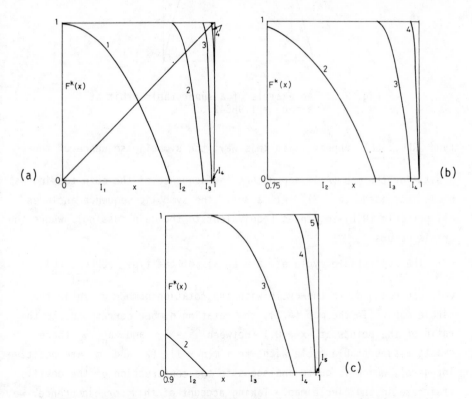

Fig. 2.32 The induced map for the logistic map (A.1).
(a) a = 1.9853, (b) a = 1.98542 (only a part),
(c) a = 1.990 (only a part).

Let us construct the symbolic dynamics by assigning R for x > 0 and L for x < 0. For $a < a_n$, the symbolic sequence of an orbit does not contain the sequence of $\underbrace{LL...LL}_{m}$ with m > n. At $a = a_n$ a superstable cycle $x_0 = 0$, $x_1 = 1$, $x_2 = 1 - a$, ..., $x_{n+1} = 0$ (see Fig. 2.33 for an example) appears. For $a > a_n$ a symbolic sequence of the

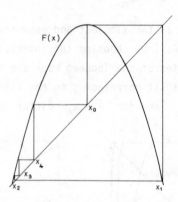

Fig. 2.33 An example of a superstable orbit at
$a = a_4 = 1.98542...$

type $\underbrace{RLL...LLR}_{n}$ appears. In this way, the symbolic sequence of the

orbits for the map (A.1) approaches the sequence of the coin tossing as n is increased (a → 2). At a = 2, the symbolic sequence includes all possible (R,L)-sequences (equivalent to the coin tossing), where the crisis occurs.

The superstable cycle at $a = a_n$ is of the type $(O R \underbrace{LL...LL}_{n-1})$,

which corresponds to the cycle with the rotation number = 1/n in the circle map. (In the map (A.1), the rotation number corresponds to the ratio of the points at x > 0.) Between $a = a_n$ and a_{n+1}, there exists a superstable cycle with kn + m(n + 1) (k and m are positive integers), which is quite analogous to the construction of the devil's staircase in the circle map. Taking account of this correspondence, we can extract the structure of the devil's staircase (Farey series) from the bifurcation sequence of all unimodal mappings. The Fibonacci- and

period-adding sequences in the BZ-map[6)-8)] can be regarded as an illustration of the above correspondence.

In the circle map, similarity and scaling of the period-adding
sequence are studied in some detail. The scaling property for the
period-adding sequence near the crisis is shown easily in the following
manner:

At $a = a_n$ and $a = a_{n+1}$ the relations

$$F^{n+1}(x = 0; a = a_n) = 0 \qquad \qquad \text{(A.2)}$$

and

$$F^{n+2}(x = 0; a = a_{n+1}) = 0 \qquad \qquad \text{(A.3)}$$

hold. Assuming that $\delta a_n = 2 - a_n$ is small, we have $(\delta a_n - \delta a_{n+1}) \propto (A_n + B_n)^{-1}$, where $A_n = \partial F^n(x = 1 - a_n, a = a_n)/\partial x$ and
$B_n = \partial F^n(x = 1 - a_n, a = a_n)/\partial a$. Using the chain rules and taking into
account the fact that a large number of periodic points are located
close to the unstable fixed point $x = -1$ for a large n, we have

$$A_n \propto (F'(x = -1, a = 2))^n, \quad B_n \propto (F'(x = -1, a = 2))^n \quad . \qquad \text{(A.4)}$$

Thus we have

$$\delta a_n \propto (F'(x = -1, a = 2))^{-n} = 4^{-n} \quad . \qquad \text{(A.5)}$$

In Fig. 2.34, δa_n vs. n is plotted for $n = 3,4,\ldots,11$, which agrees
with (A.5) quite well. In general, the scaling relation $\delta a_n \propto \alpha^{-n}$
holds, where α is the eigenvalue of an unstable periodic point which
causes the crisis at $\delta a = 0$.

We also note that the width of the parameter region where the cycle
with the period n stably exists obeys the scaling relation
$(F'(x = -1; a = 2))^{-2n}$.

The length of the interval I_n in the induced maps for $a \to 2$
(see Figs. 2.32a)-c)) is also proportional to $(F'(x = -1; a = 2))^{-n}$,
since the slope of $F^k(x)$ grows as $(F'(x = -1; a = 2))^k$. The DTD at

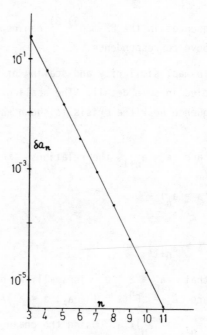

Fig. 2.34 log δa_n versus n for n = 3 to 11.
The slope of the line is -log 4.

a = 2, thus, shows the behavior

$$P(n) \propto (F'(x = -1; a = 2))^{-n} \qquad (A.6)$$

where P(n) is the ratio of the symbolic sequence $\underbrace{RLL...LLR}_{n}$, which

corresponds to the DTD in §2.9.

The similarity and scaling of the period-adding sequence are useful to characterize the property of chaotic orbits near the crisis.

Chapter 3

TRANSITION FROM TORUS TO CHAOS ACCOMPANIED BY FREQUENCY LOCKINGS WITH SYMMETRY BREAKING

Circle Limit IV
by *M. C. Escher*

§3.1 Introduction

In the previous chapter we have studied the similarity of the frequency locking using a one-dimensional circle map. When we take the Poincaré section of a torus, there remain still two variables, i.e., amplitude and phase. In the previous chapter, our interest was restricted only to the phase variable. It will be of importance, however, to study two-dimensional mappings. We use a map with a non-constant Jacobian. Particularly, we take a coupled-logistic map, that is

$$\begin{cases} x_{n+1} = 1 - Ax_n^2 + D(y_n - x_n) \\ \\ y_{n+1} = 1 - Ay_n^2 + D(x_n - y_n) \end{cases} . \qquad (3.1.1)$$

Using the well-known transformation $A = \lambda(\lambda/4 - 1/2)$, $x = (x' - 1/2)/(\lambda/4 - 1/2)$, and $y = (y' - 1/2)/(\lambda/4 - 1/2)$, Eq. (3.1.1) can be written as $x'_{n+1} = \lambda x'_n(1 - x'_n) + D(y'_n - x'_n)$ and $y'_{n+1} = \lambda y'_n(1 - y'_n) + D(x'_n - y'_n)$. Thus this map represents a system with two logistic models coupled by a linear term. It can be regarded as a simple model of two coupled systems,[1)-3)] each of which exhibits a period-doubling bifurcation route to chaos. This map may serve us to understand how coupled oscillators, such as Josephson junction or chemically reacting cells, show various behaviors, for example, entrainment, quasi-periodicity and chaos.

The map (3.1.1) is simple among maps with non-constant Jacobian. It exhibits, however, various phases, such as cycles, torus, frequency locking, chaos and hyperchaos, as will be shown later. In this chapter we focus our attention mainly on the mechanism of frequency locking at the transition from torus to chaos.

In §2, a phase diagram and some general results for the map (3.1.1) are given. Various types of the attractors (torus, chaos and hyper-chaos[4)]) are classified by the Lyapunov exponents.

In §3, the period-adding sequence of the frequency locking is studied according to the theory of a one-dimensional mapping.[5),6)] Various scalings and the similarity of the Lyapunov exponents will be checked.

In §4, we discuss the frequency locking accompanied by the symmetry breaking which was found for the map (3.1.1). After this symmetry breaking occurs, the oscillations of the two cells (i.e., x and y) become different and two types of the oscillations appear. The basin of each oscillation is also studied. It is very complicated and it will be difficult to tell which type of oscillations appear from a given initial value. Especially, two basins form a stripe structure in a self-similar manner near the line $y = x$.

The transition from torus to chaos has been observed in various two-dimensional mappings. If we restrict ourselves to the local properties of the frequency locking, the previous theory of scalings in a one-dimensional mapping[6)] is valid. However, global properties must be studied in order to understand the transition to chaos and how the dimension of the attractor grows. We discuss this problem in §5, but detailed study of the global properties is left to the future.

§3.2 Phase Diagram and General Aspects of the Coupled-Logistic Map

In this section, we give some analytic properties and the phase diagram of the map (3.1.1). Especially, the case $D = 0.1$ is studied in detail. For the case $D = 0$, the map (3.1.1) is decomposed into two independent logistic maps, which show the period-doubling bifurcations to chaos[7)] ($1 \rightarrow 2$ at $A^{(2)} = 0.75$, $2 \rightarrow 4$ at $A^{(4)} = 1.25$, $4 \rightarrow 8$ at $A^{(8)} \simeq 1.368$, $\cdots$, and $A^{(\infty)} \simeq 1.401155\cdots$). If a uniform state ($x = y$) is stable, the phase diagram is the same as the case with $D = 0$. The condition for the stability of a uniform state is given as follows.

Using the transformation $\xi_n = (x_n + y_n)/2$ and $\eta_n = (x_n - y_n)/2$, we have,[*)]

$$\xi_{n+1} = 1 - A(\xi_n^2 + \eta_n^2)$$

$$\eta_{n+1} = - 2\eta_n(A\xi_n + D) .\qquad\qquad (3.2.1)$$

Thus, η_n is given by

$$\eta_n = \prod_{k=0}^{n-1} (- 2(A\xi_k + D))\eta_0 \qquad\qquad (3.2.2)$$

The stability of the state $x_k = y_k = \xi_k$ (i.e., $\eta = 0$) is given by

$$\left| \prod_{k=1}^{p} (- 2(Ax_k + D)) \right| < 1 \qquad\qquad (3.2.3)$$

where p is the period for the logistic map and it takes infinity for chaotic states. For example, the fixed point $(x = y = (\sqrt{1 + 4A} - 1)/(2A)$ for $A < 3/4)$ is stable for $|2D - 1 + \sqrt{1 + 4A}| < 1$ and the two-cycle $(x_{1,2} = y_{1,2} = (1 \pm \sqrt{4A - 3})/(2A)$ for $A < 5/4)$ is stable for

---

*) The 3-dimensional mapping $\xi_{n+1} = F(\xi_n, \eta_{n+1}) + \zeta_n$, $\eta_{n+1} = G(\xi_n, \zeta_n)\eta_n$, $\zeta_{n+1} = b\zeta_n$ is invertible if $G(\xi_n, \zeta_n) \neq 0$, since (ξ_n, η_n, ζ_n) is represented as a set of functions of $(\xi_{n+1}, \eta_{n+1}, \zeta_{n+1})$ in that case. Our mapping (3.2.1) is considered as the overdamped limit $b \to 0$ of the invertible mapping

$$\begin{cases} \xi_{n+1} = 1 - A(\xi_n^2 + \dfrac{\eta_{n+1}^2}{4(A\xi_n + D)^2}) + \zeta_n \\[2mm] \eta_{n+1} = - 2\eta_n(A\xi_n + D) \\[2mm] \zeta_{n+1} = b\zeta_n \end{cases} \qquad (3.2.1')$$

if $A\xi_n + D \neq 0$ (i.e., $x_n \neq y_n$). Thus, the relation of the map (3.2.1') with the coupled logistic map is just like the relation of the Hénon map (1.2.4) with the logistic map (1.2.3).

$|1 + D + D^2 - A| < 1/4$. Perturbation theory for the stability of uniform states are given in Ref. 1).

We also note that from Eq. (3.2.2)

$$\prod_{k=1}^{p} (- 2(A\xi_k + D)) = 1 \qquad (3.2.4)$$

for the general p-cycle point with $x \neq y$.

Afterwards, the results for $D = 0.1$ is shown in detail. The rough phase diagram is given in Fig. 3.1. For $A < 0.56$ the fixed point is stable and the two-cycle with $x = y$ is stable for $0.86 < A < 1.25$ from the stability condition given above. There exists a two-cycle point with $x \neq y$, which is given by

$$\xi_1 = \xi_2 = (\tfrac{1}{2} - D)/A = 0.4/A$$

$$\eta_1 = - \eta_2 = \{A - [(1 - D)^2 - \tfrac{1}{4}]\}^{1/2} = (A - 0.56)^{1/2} \qquad (3.2.5)$$

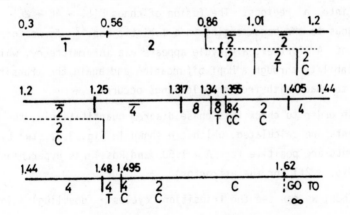

Fig. 3.1 The rough phase diagram of the coupled-logistic map 3.1.1 with $D = 0.1$. The n-cycle with $x = y$ is denoted by $\bar{n}$, while the one with $x \neq y$ is denoted by n. The 'm*torus' and n-chaos (see the text for these terminologies) are represented by mT and nC, respectively. For $0.86 \leq A \leq 1.26$, the attractors are split into two basins, which are written in two lines.

for $0.56 < A \leq 1.010$. At $A \simeq 1.010$, this cycle loses its stability via a Hopf bifurcation and a two-torus appears. (In this chapter, we use a terminology 'torus', regarding that the map is a projection onto a surface from higher-dimensional motions, and use a word 'n*torus' for the torus separated by n times (i.e., n-th iterated map gives a single torus).*) As we increase the value of A, the transition from torus to chaos with frequency locking occurs. We note that the cycle with $x = y$ (period-two for $A < 1.25$ and period-four for $A > 1.25$) and the state with $x \neq y$ coexist for $0.86 < A \leq 1.26$. At $A \simeq 1.26$, the chaos with $x \neq y$ becomes unstable and only the 4-cycle with $x = y$ remains.

The 4-cycle becomes unstable at $A \simeq 1.317$, and an 8-cycle with $x \neq y$ appears, which becomes unstable and an '8*torus' appears at $A \simeq 1.34$. The torus goes to chaos accompanied by frequency lockings. We study this transition in detail in later sections. The figures of the attractors are given in Figs. 3.2a), b) (8*torus), c) (8*chaos), d) (4*chaos) and e) (2*chaos). We use a word "n*chaos" for the chaos split into n regions. The fusion of chaos $(8C \to 4C \to 2C)$ is analogous to the band merging in a logistic map.[7] As we increase the value of A further, a 4-cycle appears via intermittency, which loses its stability through a Hopf bifurcation and again the transition from torus to chaos with frequency lockings occurs.

In order to check this phase diagram quantitatively, the Lyapunov exponents are calculated, which are shown in Fig. 3.3. Two Lyapunov exponents are positive for $A \geq 1.50$ and Rössler's hyperchaos[6] exists (see Fig. 3.2f) for the attractor).

Thus, we can see the transition "cycles → (doubling) → longer

---

*) Note that this terminology is used only in this chapter. In other chapters, m-torus means a quasiperiodic state with m incommensurate frequencies.

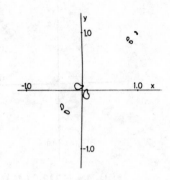

(a) A = 1.350 (8T)

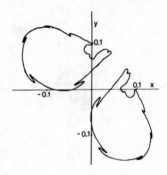

(b) A = 1.3525 (8T) (only the region $|x| < 0.2$
and $|y| < 0.2$).

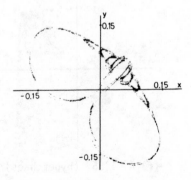

(c) A = 1.355 (8C) (only the region $|x| < 0.2$
and $|y| < 0.2$).

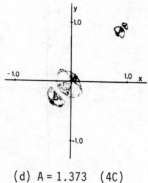

(d) A = 1.373   (4C)

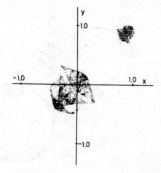

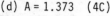

(e) A = 1.40   (2C)

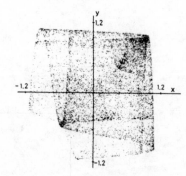

(f) A = 1.55   (hyperchaos)

Fig. 3.2   The attractor of the map (1.1) with
D = 0.1   for (a)~(f).

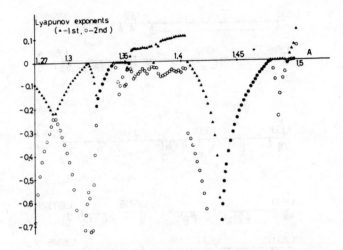

Fig. 3.3 The first and second Lyapunov exponents.
We took 200000 iterations for the
calculations.

cycle → (Hopf bifurcation) → torus → various frequency lockings → chaos
→ evolution of chaos → (fusion of chaos) → hyperchaos" in our model.

§3.3 Scaling of the Period-Adding Sequence at the Frequency Locking

In this section we study the similarity of the locking states.
Here and in what follows, we express the period by the value divided by
8 (i.e., n-cycle means 8n-cycle for the original map (3.1.1), since the
8th-iterated structure as is seen in Fig. 3.2a) is always conserved for
A ≤ 1.3555, where the transition 8C → 4C occurs. The periods of the
stable cycles which appear from A = 1.3500 to 1.3544 are given in
Figs. 3.4a), b) and c).

The sequence of the periods that is easiest to be observed in this
region is given by

$$Q_n = 8n - 1 \qquad (3.3.1)$$

with the rotation number

$$w_n = \frac{P_n}{Q_n} = \frac{2n}{(8n - 1)} \quad , \qquad (3.3.2)$$

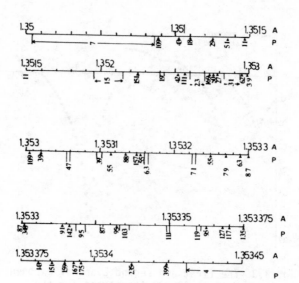

Fig. 3.4   The periods of stable cycles (lockings) which appear
          a) from A = 1.350 to 1.353 (we increase the param-
          eter A by 0.000025), b) from A = 1.353 to 1.3533
          (we increase the parameter A by 0.000005), c) from
          A = 1.35330 to 1.35345 (we increase the parameter
          A by 0.000001). We regard the attractor as p-cycle
          if (x_{400000}, y_{400000}) coincides with $(x_{400000+8p},$
          $y_{400000+8p})$ with the accuracy of 10^{-8}. The values
          of p are written below the line. If no value is
          written, there is no cycle with the period shorter
          than 2500.

where P_n is the number of the rotations around the center of a torus
during Q_n cycles. We define A_n by the value at which the Q_n-cycle
appears. As n goes to infinity, the rotation number converges to 1/4
and A_n to A_∞ = 1.35343075$\cdots$, after which the stable 4-cycle appears.
We have plotted $\log(A_\infty - A_n)$ vs log n in Fig. 3.5, which shows

$$A_\infty - A_n = 0.0148 \times n^{-2}$$

(3.3.3)

as n goes large. The width $\Delta A_n \equiv A_n^f - A_n$ (A_n^f is the value of A,
where the Q_n-cycle loses its stability) obeys the scaling $\Delta A_n \propto n^{-3}$
(see Fig. 3.5). We have also studied the distance of nearest periodic

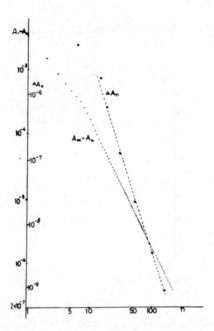

Fig. 3.5 Scaling properties of the period-adding
sequence ((8n - 1)-sequence). The quan-
tities $(A_\infty - A_n)(\cdot)$ and $A_n(\blacksquare)$ vs n
are plotted.

points ΔR_n and verified $\Delta R_n \propto n^{-2}$. Thus, the critical exponent is
the same as the one-dimensional case in Chap. 2[5),6)].

Next, we study to which category in §2.5 this sequence belongs.
The local period-doubling bifurcation occurs for $n \geq 5$ (e.g., 39 → 78
→ 156 → 312 →...→ chaos), which obeys the theory of Feigenbaum. Thus,
this sequence belongs to the Case III of §2.5. The scaled first Lyapunov
exponent ($n\lambda_n$ as a function of $n^3(A - A_n)$) is given in Fig. 3.6, which
seems to assure the existence of the fixed point function.[6)] We note the
similarity between Fig. 3.6 and Fig. 2.10c) or Fig. 2.13. The second
Lyapunov exponent is large in magnitude compared with the first and the
variation of it is small. The application of the theory of the one-
dimensional mapping, therefore, seems to be justified.

Thus, the numerical results reproduce the one-dimensional map

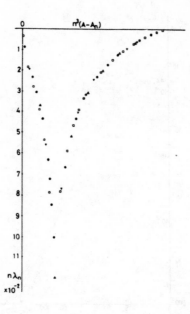

Fig. 3.6 The similarity of the first Lyapunov exponent
λ_n. The Lyapunov exponent is scaled by n^{-1},
while $(A - A_n)$ is scaled by n^{-3}, $\bullet$, $\triangle$, $\blacktriangledown$,
and $\square$ show the scaled exponents for $n = 14$,
18, 30 and 101, respectively.

theory in Chap. 2.

§3.4 Frequency Locking with Symmetry Breaking

According to the theory based on the map

$$\theta_{n+1} = \theta_n + A \sin(2\pi\theta_n) + D \quad (\text{mod } 1) \quad , \qquad (3.4.1)$$

there is a sequence with the rotation number $(qn + s)/(pn + r)$ ($n =$
$1,2,\cdots\cdots$) between the lockings with q/p and s/r. In this case p
and q are relatively prime. The sequence with the rotation number
$2n/(8n - 1)$ in §3, however, does not satisfy this property. We also
have to note that the period of the locking which appears at A_∞ is not
8 but 4. What causes this difference?

First, we note that two types of 4-cycles (32-cycle for the original map) coexist. The basin for the attraction breaks into two parts. The attractor is symmetric about $y = x$ for $A < A_\infty$, but each 4-cycle at $A > A_\infty$ is not symmetric about $y = x$ (see Fig. 3.7). One type of the 4-cycle is a mirror image of the other 4-cycle about $y = x$. Thus, the symmetry breaking occurs at $A = A_\infty$.

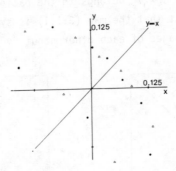

Fig. 3.7 Two types of the 4-cycles (32-cycle for the original map) at $A = 1.35344$. ● and △ denote the two 4-cycles respectively. We have plotted only 8 points nearest to the origin. We note that the cycle △ is a mirror image of the cycle ● about $y = x$.

For $A < A_\infty$, the cycle (or chaos) consists of the two nearly-four-cycle oscillations (i.e., points near one type of the 4-cycle at $A = A_\infty$ and points near the other type) and the transition between the two regions. The time for the transition grows by $(A_\infty - A)^{-1/2}$ (tangent bifurcation mechanism), which diverges at $A = A_\infty$, and the symmetry is broken for $A > A_\infty$. The locking near $A < A_\infty$ (e.g., $(8n-1)$-sequence in §3) is made of the points close to the 4-cycles at $A > A_\infty$ (● and △ in Fig. 3.7) and of the points between these points. The number of the former points increases as A approaches A_∞ (it increases in proportion to n for the $(8n-1)$-sequence), while the number of the latter does not.

Thus, the period-adding sequence is not $(4n - 1)$ but $(8n - 1)$. In general, we can conclude that the period-adding sequence with the

98

rotation number $(2qn + s)/(2pn + r)$ appears before the locking q/p, which breaks some two-fold symmetry and has two basins.

Next, we study the structure of basins. In Figs. 3.8a) and b), it is shown which type of the 4-cycles appears from a given initial value (x_0, y_0) at $A = 1.35344$. The figures are antisymmetric about $y = x$, in the sense that if the point (x_0, y_0) belongs to the basin for one 4-cycle, the point (y_0, x_0) belongs to the basin for the other 4-cycle. This is due to the fact that the map (3.1.1) is symmetric and the two 4-cycles are mirror images of each other about $y = x$.

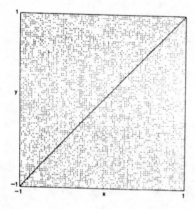 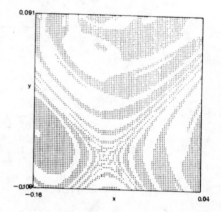

(a) We have studied the map 3.1.1 with the initial values $(x_0, y_0) =$ $(-1.0 + 2i/100, -1.0 + 2j/100)$ $(i, j = 1, 2, \ldots, 100)$. If the point $(x_0(i), y_0(j))$ is a basin for the 4-cycle of the type ● of Fig. 4, we have put a dot, while we have not put a dot, if the point is a basin for the 4-cycle of the other type.

(b) We have increased the resolution of Fig. 8(a). We have studied 100 × 100 points in $-0.16 \leq x_0 \leq 0.04$ and $-0.109 \leq y_0 \leq 0.091$. In this region the 4-cycle of Fig. 7 exists. We note the stripe structure near the line $y = x$.

Fig. 3.8 The basin for the attraction for each 4-cycle at $A = 1.35344$.

First, we note that the structure is very complicated. This will be due to the effect of stretching and folding which appears only in the transient phenomenon for this value of A, and which causes chaos for larger A. Near the line y = x the basins form a self-similar structure (see Fig. 3.9). Around this value of A, there exists a periodic

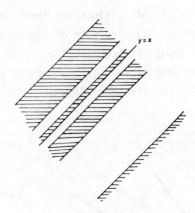

Fig. 3.9 The stripe structure of the basins around y = x. The shaded region is a basin for one 4-cycle, while the blank region is a basin for the other 4-cycle. When we enlarge the figure and increase the resolution by $\alpha^2 \simeq 3.0$, the same pattern appears.
(The scale of this figure is arbitrary in this sense.)

saddle of period-4 on the line y = x (which is a 4-cycle for the single logistic map; see §2). The matrix of the fourth iteration of the map (3.1.1) linearized around this saddle has an eigenvalue $\alpha \approx -1.73$ (< -1). Thus, a zone of a basin at x < y is reduced as $|\alpha|^{-1}$ and appears at x > y (because α is negative). This process is repeated infinitely and makes a self-similar structure of the basin near y = x. The value of the scaling factor $|\alpha|^{-1}$ is verified through numerical results (see Fig. 3.9). We can construct an example of a self-similar structure of basins by using some one-dimensional mappings.[8]

It is interesting to regard that x and y represent the observables of the two cells in a coupled system. Then the frequency locking with symmetry breaking of this section can be regarded as the appearance of a spatio-temporal order. At $A > A_\infty$, the oscillations of the two cells become completely different.

The global structure of lockings are investigated in Chap. 2. In our problem here, the global structure of lockings is more complex, since there are two types of lockings, i.e., with or without symmetry. The rough phase diagram of frequency lockings is given in Fig. 3.10, where the number "n" denotes a locking of n-cycle (8n-cycle for the original map).

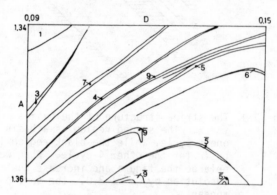

Fig. 3.10 The frequency lockings around the transition from torus to chaos. The number (1~9) is the period of the stable cycle. We have plotted only cycles with the period shorter than 10. The n-cycle (8n-cycle for the original map) which appears after the fusion 8C → 4C is represented by ñ, which is better to be regarded as 4 × 2n cycle.

First, we note that there is a period-adding sequence 3 → 4 → 5 → 6 → (→ 7 → ...). These cycles break the symmetry, that is, the attractor is not symmetric about y = x and there exist two types of cycles as is the case for the period 4. There appear lockings with the period 7 (between 3 and 4), 9 (between 4 and 5), 11 (between 5 and 6), etc., which is symmetric about y = x. The (8n-1)-sequence, studied in

previous sections, appears between the periods 7 and 4. The lockings
with longer periods are constructed in a way similar to the case for
one-dimensional mapping (i), (ii), (iii), ...). As is seen in Fig. 3.10,
the locking with symmetry breaking (e.g., 3 or 4 or 5 ...) occurs at
various values of A and D.

The window ($\bar{5}$ in Fig. 3.10) goes to chaos by the period-doubling
bifurcation, as the parameter A is increased. The attractor of this
chaos is not symmetric about $y = x$ (see Fig. 3.11a)). As the parameter A is increased further, the strange attractor with symmetry
appears again (see Fig. 3.11b)). Thus, the chaos-chaos transition with
symmetry breaking is also observed in our system.

§3.5 Discussion

Transition from torus to chaos accompanied by frequency lockings
have frequently been observed in dissipative two-dimensional mappings.
The scenario is as follows:

i) Torus appears via a Hopf bifurcation.

ii) The shape of torus is distorted (see Fig. 3.2b)).

iii) As the process ii) goes on, the regions of frequency lockings
increase. They form a structure of devil's staircase.

iv) Chaos appears through a period-doubling or a tangent bifurcation of some frequency-locked cycle at some value of the
bifurcation parameter.

v) The dimension of the attractor increases, which is seen in
the increase of two Lyapunov exponents.

vi) As we increase the bifurcation parameter, the sum of two
Lyapunov exponents become positive and the attractor covers
a two-dimensional area (see Fig. 3.2e)).

vii) Two Lyapunov exponents take positive values (hyperchaos).

This process (i) → vii); it may stop at some step and the inverse
process or transition from chaos to cycle can occur) was observed in

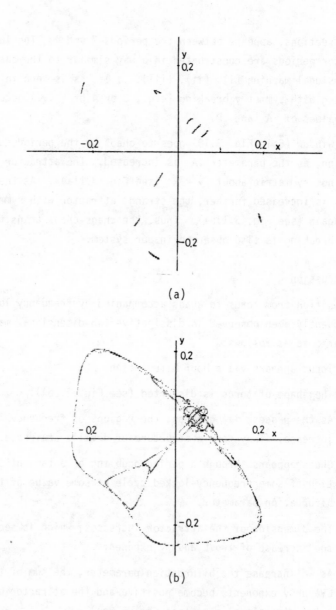

(a)

(b)

Fig. 3.11 The attractor of the map 3.1.1 with
D = 0.14 for (a) A = 1.35704, (only
one of the two types of the attractors); (b) A = 1.35710.
The chaos is of the type "4C" in our
notation and only the region nearest
to the origin is shown in these figures.

our map (3.1.1) in various parameter regions and was also observed for
other two-dimensional mappings, such as

$$
\begin{cases}
r_{n+1} = \gamma r_n - g r_n^3 + \tilde{A} \, r_n \sin (2\pi\theta_n) \\
\theta_{n+1} = \theta_n + \alpha r_n^2 \quad (\text{mod } 1) \quad .
\end{cases}
\tag{3.5.1}
$$

The distortion of torus ii) is always seen in these examples, which
will be treated in the next chapter in detail.

At the frequency lockings iii), it is easy to observe the period-
adding sequence. The scaling properties given in the previous chapter
hold in these examples.

Our mapping (3.1.1) has a symmetry under the exchange of variables
$x \leftrightarrow y$. It is reflected in the properties of the attractor. The
attractors shown in Fig. 3.1 (such as m, mT, C, mC: (m = 2,4,8))
are symmetric about $y = x$. The symmetry breaking occurs as a locking
or as a chaos-chaos transition. The critical phenomenon near this
locking is ruled by a tangent bifurcation. In this case, the period
of the period-adding sequence with similarity is not (pn + q) but
(2pn + q) (p is a period of the cycle which appears at $n \to \infty$). The
symmetry breaking of this type has not yet been studied well. It may
give a new insight upon the onset of a spatio-temporal order.

The basin of each cycle with broken symmetry is very complicated,
as was shown in Figs. 3.8a) and b). Thus, it will be rather difficult
to predict the final state from a given initial value. The complicated
structure is typically seen near the saddle line $(y = x)$ as a self-
similar stripe structure. This structure is expected to appear near
unstable cycles, if one stripe structure of a basin exists. Thus, this
structure of a basin may be observed in various nonlinear systems.

Fractal properties of basin structures have recently been investigated by Grebogi, Ott, Yorke[9],[10] and by Takesue and the author.[8]
Grebogi et al.'s work have elucidated the fractal basin boundary and
final state sensitivity in such systems.

After the work in this chapter was published, some studies on the coupled logistic maps have appeared, including the effect of noise,[11] the case with nonlinear coupling,[12] a feature of coupled chaos,[13] and the method how the coupled maps are obtained from the coupled differential equations[14] and search for an N-tupling bifurcation.[15] Coupled maps will be used to study the features of various systems.

It is also of importance, of course, to study N-coupled logistic maps for N > 2. This problem will be investigated in Chap. 7.

REFERENCES

*) The contents of this chapter are essentially based on K. Kaneko, Prog. Theor. Phys. 69 (1983) 1427-1442.

1. H. Fujisaka and T. Yamada, Prog. Theor. Phys. 69 (1983) 32.

2. I. Schreiber and M. Marek, Phys. Lett. 91A (1982) 263.

3. M. Sano and Y. Sawada, Phys. Lett. 97A (1983) 73.

4. O.E. Rössler, Phys. Lett. 71A (1979) 155.

5. K. Kaneko, Prog. Theor. Phys. 68 (1982) 669.

6. K. Kaneko, Prog. Theor. Phys. 69 (1983) 403.

7. P. Collet and J.P. Eckmann, Iterated Maps on the Intervals as Dynamical Systems (Birkhäuser, 1980).

8. S. Takesue and K. Kaneko, Prog. Theor. Phys. 71 (1984) 35.

9. C. Grebogi, E. Ott, and J.A. Yorke, Physica 7D (1983) 181.

10. C. Grebogi, S.W. McDonald, E. Ott, and J.A. Yorke, Phys. Lett. 99A (1983) 415.

11. T. Hogg and B.A. Huberman, Phys. Rev. A29 (1983) 295.

12. J.M. Yuan, M. Tung, D.H. Feng, L.M. Narducci, Phys. Rev. A28 (1983) 1662.

13. K. Tomita and Lee, private communication.

14. T. Yamada and H. Fujisaka, Prog. Theor. Phys. 70 (1983) 1240.

15. J. Frøyland, Physica 8D (1983) 423.

Chapter 4

OSCILLATION AND FRACTALIZATION OF TORI

Rippled Surface
by *M. C. Escher*

§4.1 Introduction

Many two-dimensional mappings show the transition from torus to chaos. One characteristic feature near the transition point is the oscillatory behavior of torus. It is related to the instability along the amplitude direction. In the experiments of Rayleigh-Bénard convection, the oscillatory behavior of tori has recently been observed. The attractors in the Rayleigh-Bénard experiment by Bergé[1] and M. Sano[2] show the oscillatory behavior of tori (see Fig. 4.1). In §4.2, the

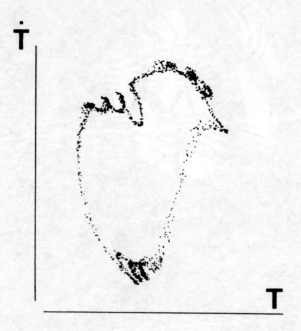

Fig. 4.1 Experimental Poincaré sections corresponding to the phase space trajectories in unsteady Rayleigh-Bénard convection. The test fluid is silicone oil of Prandtl number 130, confined in a rectangular geometry of horizontal extension 2d × 1.2d, d = 2 cm being the height of the layer. A three-dimensional phase space is defined from 3 independent dynamical variables, which are obtained from the deflection of light beams, due to time dependent thermal gradients. (cited from Ref. 1)

oscillation of torus will be investigated using a simple two-dimensional mapping (delayed logistic model). The oscillation of torus will be related to the damped oscillation of an unstable manifold of a periodic saddle.

In usual two-dimensional mappings, however, the region of lockings increases as the nonlinearity parameter is increased. Thus, it will be rather difficult to study the oscillatory behavior in detail. In §4.3, the modulation map is introduced, which has no locking region by its construction. As the bifurcation parameter is increased, the oscillation becomes stronger and stronger, till the torus collapses at some critical value of the parameter. The torus loses its differentiability and seems to become fractal at the critical point. Fractal dimension at the onset of chaos is numerically calculated. Critical phenomena of "fractal torus" will be investigated using a functional mapping, which is derived from the equation of the invariant curve. Universality of the fractalization phenomena is also studied. The functional mapping itself involves new and interesting problems. Discussions will be given in §4.4, where these problems are also considered.

§4.2 Oscillation of Torus in Two-Dimensional Mappings

In various two-dimensional mappings, oscillation of torus appears before the transition from torus to chaos undergoes. In the present section, we give a numerical result for a typical mapping and discuss the mechanism of the oscillation. The map we investigate here is a two-point delayed logistic map,

$$\begin{cases} x_{n+1} = Ax_n + (1 - A)(1 - Dy_n^2) \\ \\ y_{n+1} = x_n \end{cases} \tag{4.2.1}$$

(see Appendix for the meaning of the mapping). The fixed point $x = y = (\sqrt{1 + 4D} - 1)/(2D)$ loses its stability at $D = 1/(1 - A)$ via a Hopf bifurcation and a torus appears for $D > 1/(1 - A)$. As D is increased further, the transition to chaos occurs accompanied by various frequency lockings.

The attractors are given in Figs. 4.2a)-g). Thus, the transition proceeds as follows:

i) oscillation of torus (a), b))

ii) various kinds of lockings appear (which form the devil's stair-case) (c))

iii) chaos emerges from a locking (via period-doubling or inter-mittency) (d))

iv) the width of the belt-like attractor along the amplitude direction increases (the dimension of the attractor becomes two) (e), f))

v) the unstable fixed point $x = y = (\sqrt{1 + 4D} - 1)/(2D)$ becomes a snap-back repeller[3] (g)).

Next, we study the oscillation of torus in more detail. For simplicity, we consider the map (4.2.1) for smaller values of A. In this case a locking to a 4-cycle is dominant near the transition to chaos. The oscillation of torus is typically seen near this locking (see Fig. 4.3). The magnified figures of the attractors are given in Fig. 4.4a) (torus), 4.4b) (locking) and 4.4c) (chaos).

The mechanism of this oscillation is understood as follows: We consider a locking (4-cycle) which appears at a different but close value of the bifurcation parameter. When the locking occurs, periodic saddles (with period 4) also appear. If the unstable manifold of a periodic saddle crosses to a stable manifold of a stable cycle, it must cross infinite times. We investigate the Jacobi matrix $J(x_j^{s(u)}, y_j^{s(u)})$, where $(x_j^{s(u)}, y_j^{s(u)})$ is the stable (unstable) periodic point $(j = 1,2,3,4)$. The matrix $[J(x_1^s, y_1^s)J(x_2^s, y_2^s)J(x_3^s, y_3^s)J(x_4^s, y_4^s)]$ has the eigenvalues $\lambda_{sp}(0 < \lambda_{sp} < 1)$ and $\lambda_{sa}(-1 < \lambda_{sa} < 0)$, while the matrix $[J(x_1^u, y_1^u)J(x_2^u, y_2^u)J(x_3^u, y_3^u)J(x_4^u, y_4^u)]$ has the eigenvalues $\lambda_{up}(\lambda_{up} > 1)$ and $\lambda_{ua}(-1 < \lambda_{ua} < 0)$ ("p" and "a" represent the "phase" and "amplitude"). The manifolds corresponding to these eigenvalues are schematically given in Fig. 4.5. If the manifolds M_{up} and M_{sp} intersect transversally at P, they intersect also at the points

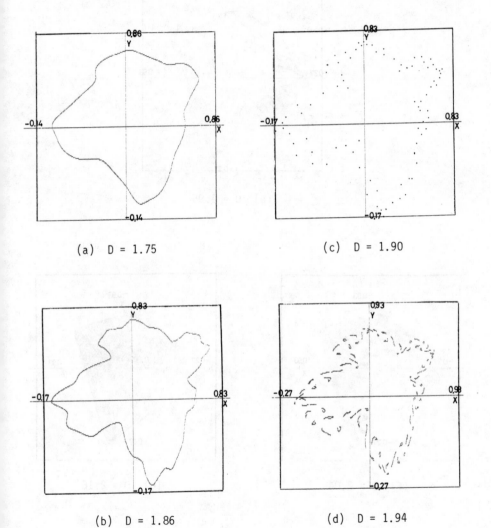

(a) D = 1.75

(c) D = 1.90

(b) D = 1.86

(d) D = 1.94

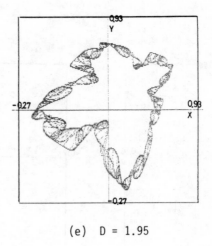

(e)   D = 1.95

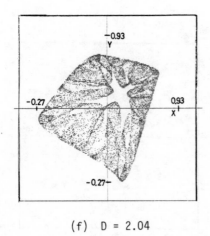

(f)   D = 2.04

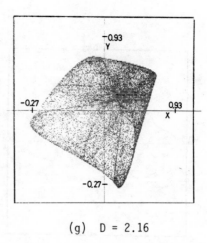

(g)   D = 2.16

Fig. 4.2   Attractor of the delayed logistic map
(4.2.1), with   A = 0.3.

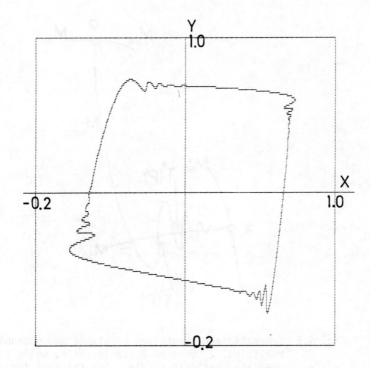

Fig. 4.3 Attractor of the map (4.2.1) with A = 0.12 and D = 1.3.

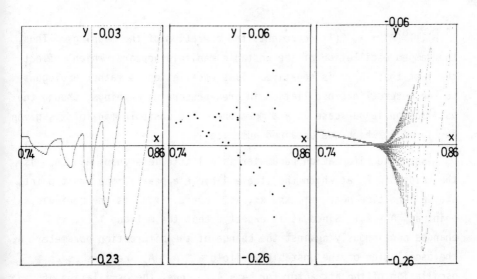

Fig. 4.4 A part of the attractor of the map (4.2.1) with A = 0.12.

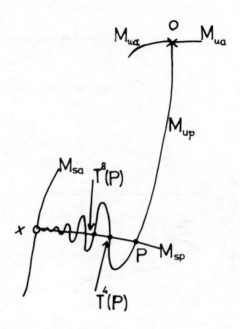

Fig. 4.5 Schematic representation of stable and unstable
manifolds. ○ denotes a stable cycle, while X
denotes a periodic saddle. See the text for
other notations.

$T^4(P)$, $T^8(P)$,···, (T represents the operation of the mapping). Thus,
the damped oscillation of the unstable manifold appears, which reflects
the fact that λ_{sa} is negative. This oscillation is rather analogous
to the heteroclinic oscillation in area-preserving mappings, though the
oscillation in our case is a damped one. The unstable manifold is given
in Fig. 4.6, which was obtained numerically.

When the bifurcation parameter is a little less than A_4 (A_4 is
the value of A at which the stable 4-cycle appears), the orbit points
stay a long time near x_1^*, x_2^*, x_3^*, x_4^*, where $\{x_i^*\}$ is the periodic
point at $A = A_4$. Since it is expected that the motion $T^4(x, y)$
changes continuously against the change of the bifurcation parameter A,
the oscillation of the unstable manifold at $A \gtrsim A_4$ remains as an
oscillation of the attractor for $A \lesssim A_4$. Thus, the oscillation of

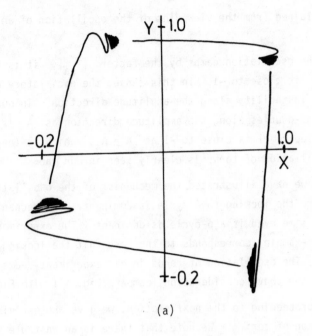

(a)

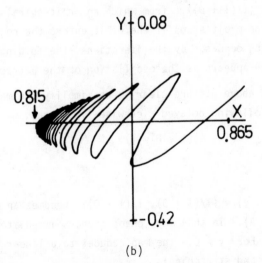

(b)

Fig. 4.6   Unstable manifolds of periodic saddles
            for the map (4.1.2) with A = 0.12 and
            D = 1.37, which are obtained numerically.
            b) enlargement of a).

torus is explained from the viewpoint of the oscillation of an unstable manifold.

Since the oscillation damps by the factor $|\lambda_{sa}|$, it is remarkably seen if λ_{sa} is close to -1. In this sense, the oscillatory behavior reflects the instability along the amplitude direction. In our map, the 4-cycle period-doubles along the amplitude direction as A is increased. Thus, the value λ_{sa} is close to -1 at $A = A_4$, which is the reason that the oscillation of torus is clearly seen in our map.

So far, we have illustrated the mechanism of the oscillation using the case where the locking to 4-cycle is dominant. The mechanism, however, is the same even if a p-cycle is dominant. The experiment by Bergé,[1] for example, corresponds to the case that the locking to 3-cycle is dominant. The oscillation of torus in his experiment seems to be well explained by the above consideration (compare Fig. 4.1 with Fig. 4.3).

Before proceeding to the next section, we give another viewpoint on the oscillation of tori.[4] We note that there is an unstable focus (in our map $x = y = 1/(1 + D)$), from which an orbit spirals out. The stretching of the orbit stops if the orbit enters the region $y < 0$, where the folding occurs. By the iterations, the folding is rotated and stretched, which appears as the oscillation of the attractor.

To see the above picture clearly, we simplify the map (4.2.1) and introduce the following delayed piecewise-linear map,

$$\begin{cases} x_{n+1} = Ax_n + (1 - A)(1 - D|y_n|) \\ y_{n+1} = x_n \end{cases}$$

(4.2.2)

Fixed point $(x, y) = (1/(1 + D), 1/(1 + D))$ becomes an unstable focus for $D > 1/(1 - A)$. In this map, chaos appears immediately for $D > 1/(1 - A)$. For $y > 0$, the map reduces to a linear transformation (i.e., rotation and stretching)

$$\begin{pmatrix} x'_{n+1} \\ y'_{n+1} \end{pmatrix} = \begin{pmatrix} A & -(1 - A)D \\ 1 & 0 \end{pmatrix} \begin{pmatrix} x'_n \\ y'_n \end{pmatrix},$$

(4.2.3)

where $x' = x - 1/(1 + D)$, $y' = y - 1/(1 + D)$. For $y < 0$, folding ($y \leftrightarrow -y$) occurs by the term $D|y|$.[*) The attractors of the map (4.2.2) are shown in Fig. 4.7. The oscillatory behavior of the attractor is clearly seen, which can be explained by the above mechanism, i.e., rotation (the eigenvalues of the matrix (4.2.3) are complex), stretching (the absolute values of them are larger than one), and folding.

The Jacobi matrix for the map (4.2.2) has the eigenvalues $(A \pm \sqrt{4c - A^2}i)$ for $y > 0$ and $(A \pm \sqrt{4c + A^2})$ for $y < 0$, where c is given by $(1 - A)D$. Since the measure for $y < 0$ is small near the onset of chaos, we can neglect the contribution from the points $y < 0$ to the lowest approximation. Then the Lyapunov exponents are given by $(1/2) \log c$ (first and second exponents are degenerate), which agree with the numerical results within 10^{-3}. Within the above approximation the rotation number θ is given by $(1/2\pi)\arccos(A/(2\sqrt{c}))$.

Thus, the Lyapunov exponents behave as L_1, $L_2 \propto \varepsilon$ near the onset of chaos, where ε denotes $D - D_c$. The measure of the points for $y < 0$, which corresponds to the disordering ratio in the circle map, is estimated by a self-consistent argument, which gives disordering ratio $\propto \varepsilon^{1/3}$, near the onset of chaos.

The oscillation near the onset of chaos $(D > 1/(1 - A))$ shows a small scale structure as is seen in Fig. 4.7 and it will be possible and interesting to study the critical behaviors of the oscillations near the onset of chaos, using, for example, the continued fraction expansion method for a given irrational rotation number.

When the rotation number is close to a simple rational value, a phenomenon similar to the band splitting in a logistic map occurs as is illustrated in Fig. 4.7e), where the rotation number is close to 10/41. The linear map (4.2.2) has abundant new features of chaos and oscillations which have to be illuminated in future.

[*)] In the region $y < 0$, the direction of rotation is almost parallel to the x-axis. Thus, the folding occurs nearly parallel to the amplitude direction.

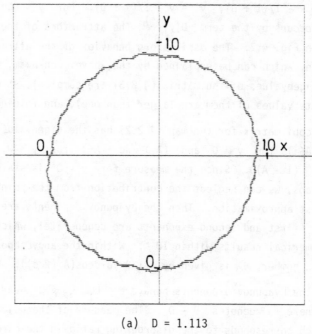

(a) D = 1.113

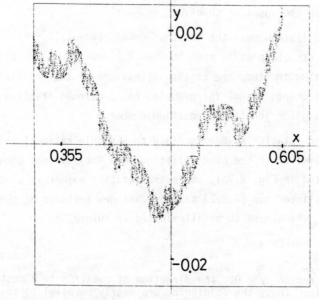

(b) D = 1.113 (only a part)

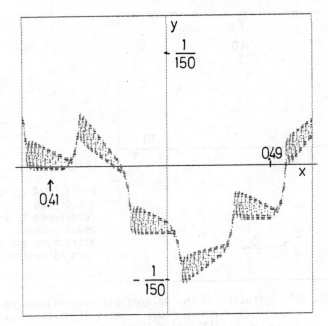

(c)   D = 1.1118 (only a part)

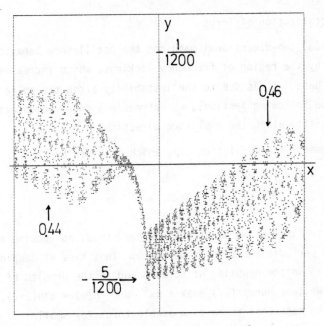

(d)   D = 1.1115 (only a part)

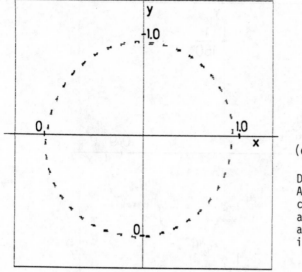

(e)

D = 1.0834 ... and
A = 0.0766 ..., which
correspond to c = 1.0008
and θ = 10/41. The
attractors are split
into 41 regions.

Fig. 4.7 Attractor of the delayed piecewise-linear map
(4.2.2) with A = 0.1 (onset of chaos occurs
at D = 1/(1 - A) = 10/9).

§4.3 Fractalization of Torus

In usual two-dimensional mappings the oscillatory behavior of torus
is masked by the region of frequency lockings, which increases near the
onset of chaos. It is due to the instability along the phase motion.
In this and following sections, we introduce a model which extracts only
the instability along the amplitude direction.

The model is a modulation map, which is given by

$$\begin{cases} x_{n+1} = f(x_n) + \varepsilon h(\theta_n) \\ \theta_{n+1} = \theta_n + c \quad (\text{mod } 1) \end{cases}$$

(4.3.1)

where c is fixed at an irrational number, thus, no locking appears and
$h(\theta)$ is a periodic function of period 1. This type of mapping has also
been studied in the doubling of torus[5] and in the problem of three-
torus.[6] We take here $f(x) = ax + bx^2$ and $h(\theta) = \sin(2\pi\theta)$, and
consider the case that there is a stable torus for small ε. For $\varepsilon \simeq 0$,
the torus is almost straight. As ε is increased, it oscillates more

and more strongly till it collapses at some critical value $\varepsilon = \varepsilon_c$ and chaos emerges (see Fig. 4.8a), b), and c) for the attractors). The value ε_c is numerically confirmed by the calculations of Lyapunov exponents.

In order to investigate the oscillation of torus in more detail, we study the equation for the invariant curve. The invariant curve $x = g(\theta)$, if it exists, must obey the functional equation

$$g(\theta + c(\mathrm{mod}\ 1)) = f(g(\theta)) + \varepsilon h(\theta) \ . \tag{4.3.2}$$

If $f(x)$ is linear (i.e., $b = 0$), Eq. (4.3.2) is solved to give

$$g(\theta) = \frac{\varepsilon(\sin 2\pi(\theta - c) - a \sin 2\pi\theta)}{\{(1 - a \cos 2\pi c)^2 + a^2 \sin^2 2\pi c\}} \ . \tag{4.3.3}$$

When $f(x)$ is nonlinear, it is difficult to obtain the analytic solution of Eq. (4.3.2). Here we search for the solution of (4.3.2) numerically by iterating the functional mapping

$$g_{n+1}(\theta + c(\mathrm{mod}\ 1)) = f(g_n(\theta)) + \varepsilon h(\theta) \ . \tag{4.3.4}$$

As a numerical technique, we replace c by a rational value c_k using a continued fraction expansion

$$c_k = 1/\{n_1 + \{1/n_2 + \{1/n_3 + \cdots + 1/n_k\}\cdots\} \ . \tag{4.3.5}$$

We study mainly the case $c = (\sqrt{5} - 1)/2$, when c_k is given by F_{k-1}/F_k, where F_k is the Fibonacci sequence. Thus, the functional map (4.3.4) is replaced by the F_k-dimensional mapping. The convergence of the iteration (4.3.4) becomes slower and slower as ε approaches ε_c, and no convergence is obtained for $\varepsilon \gtrsim \varepsilon_c$ within our numbers of iterations (5000).

The figures of the attractors tell us that the torus seems to be fractal[7] at $\varepsilon \sim \varepsilon_c$. To confirm this property, we measured the length of torus by changing the scales, i.e., we calculated the following quantities

$$L(j) = \frac{1}{j} \sum_{i=1}^{F_k} \left[\left\{ g(\frac{i + j}{F_k} (\mathrm{mod}\ 1)) - g(\frac{i}{F_k}) \right\}^2 + (\frac{j}{F_k})^2 \right]^{1/2} \tag{4.3.6}$$

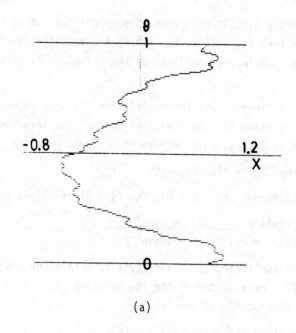

(a)

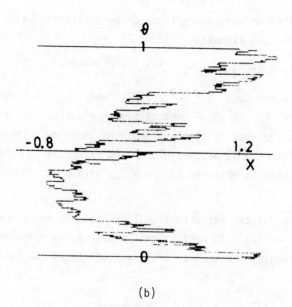

(b)

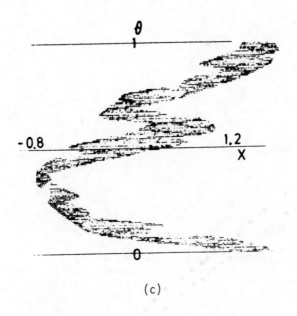

(c)

Fig. 4.8 Attractor of the map (4.3.1) with $f(x) = -x + x^2$
and $h(\theta) = \sin(2\pi\theta)$ and $c = (\sqrt{5} - 1)/2$. The
values of ε are 0.36 (torus) for a), 0.46 (torus)
for b) and 0.49 (chaos) for c).

If $L(j) \propto j^{-\alpha}$ for small j, the torus is fractal with the dimension
$d_F = 1 + \alpha$.[7] The log-log plot of $L(j)$ vs j is shown in Fig. 4.9,
which was obtained from numerical iterations of the map (4.3.4) with
$\varepsilon = 0.472 \simeq \varepsilon_c$ and $F_k = 28657$ or 46368. As is seen from this figure,
the torus at $\varepsilon \simeq \varepsilon_c$ is fractal with the fractal dimension $d_F = 1.77 \pm 0.04$.

We also counted the number of extremum points N_k by changing F_k,
by calculating the number of the integers j which satisfy
$\{g((j + 1)/F_k) - g(j/F_k)\} \times \{g(j/F_k) - g((j - 1)/F_k)\} < 0 \ (1 \leq j \leq F_k)$.
For $\varepsilon \ll \varepsilon_c$, the number N_k approaches a constant as F_k is increased,
while it goes to infinity, showing the behavior $N_k \propto F_k$ for $\varepsilon \simeq \varepsilon_c$.
Thus, there is a self-similarity at $\varepsilon \simeq \varepsilon_c$. This result also confirms
the fractal property of the torus at the onset of chaos (see Fig. 4.10).

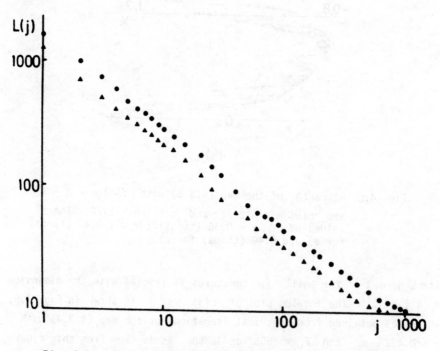

Fig. 4.9 Log-log plot of the length L(j). The value of
ε is 0.472 and F_k is 28657 (▲) and 46368 (●).
The function f(x) is given by f(x) = - x + x^2
and h(θ) = sin(2πθ).

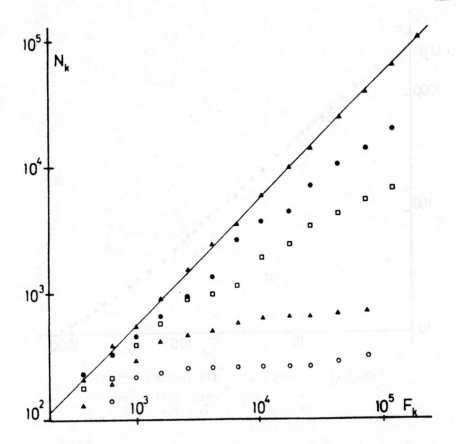

Fig. 4.10 The number of extrema N_k as a function of
F_k, for $f(x) = -x + x^2$. The values of ε
are 0.45 (○), 0.46 (△), 0.465 (□), 0.47 (●)
and 0.472 (▲).

As another example, we consider the case $c = \sqrt{2} - 1$, where the
continued fraction expansion is given by $c = 1/(2 + 1/(2 + 1/(2 + \cdots)$
$\cdots)$. Thus, the rational approximation for c is written by $c_k = G_{k-1}/G_k$ where $G_{k+1} = 2G_k + G_{k-1}$. Using this approximation, the mapping
(4.3.4) is reduced to a G_k-dimensional mapping. The log-log plot of
$L(j)$ at the onset of chaos is shown in Fig. 4.11. As is seen from this
figure, the torus becomes fractal at the onset of chaos, but its dimension

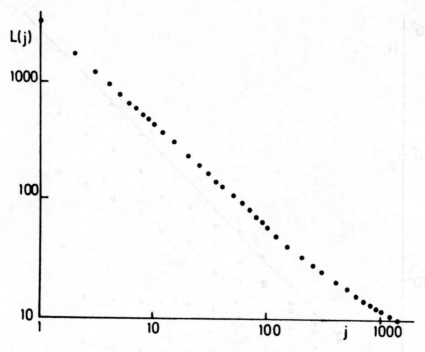

Fig. 4.11 Log-log plot of the length $L(j)$ for
$f(x) = -x + x^2$ and $h(\theta) = \sin(2\pi\theta)$.
The value of ε is 0.55417 and $c_k =$
13860/33461.

(1.85 ± 0.03) differs from the case $c = (\sqrt{5} - 1)/2$. Thus, the dimension depends on the form of the expansion of an irrational number. We also note that the critical value of the coupling ε_c for $c = \sqrt{2} - 1$ is smaller than the value for $c = (\sqrt{5} - 1)/2$ (i.e., torus with $c = (\sqrt{5} - 1)/2$ is not the "last KAM").

To study the universality of fractalization of torus, we consider the case $f(x) = x - 0.2\sin(2\pi x)$. In this case also fractalization phenomena occur as ε is increased. We calculated again the length of torus at the onset point of chaos, using the continued fraction expansions. Log-log plot of $L(j)$ is shown in Fig. 4.12, where c_k is 28657/46368 ($\approx (\sqrt{5} - 1)/2$) and 13860/33461 ($\approx \sqrt{2} - 1$), respectively. The fractal

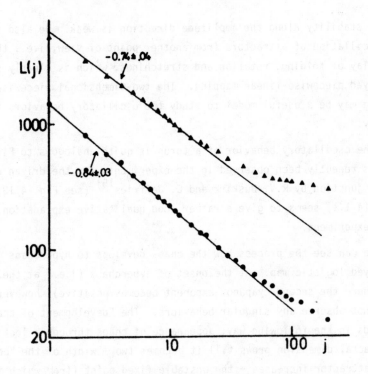

Fig. 4.12 Log-log plot of the length $L(j)$ for
$f(x) = x - 0.2 \sin(2\pi x)$ and $h(\theta) = \sin(2\pi\theta)$. The values of ε are
0.6644 ($\blacktriangle$) and 0.6837 ($\bullet$), where c_k
is given by $F_{k-1}/F_k = 28657/46368$ ($\blacktriangle$)
and $G_{k-1}/G_k = 13860/33461$ ($\bullet$),
respectively.

dimensions seem to agree with the values for $f(x) = ax + bx^2$ respectively. The detailed study on the universality, however, is left to the
future, since the accuracy is not good enough to confirm the universality.

§4.4 Summary and Discussion

In this chapter we have investigated the oscillatory instabilities
of the torus motion. First, we relate the oscillation of torus to the
oscillation of unstable manifolds. The oscillation can be typically seen

if the stability along the amplitude direction is weak. We also studied the oscillation of attractors from another point of view, i.e., the interplay of folding, rotation and stretchings, which is clearly seen in a delayed piecewise-linear mapping. The two-dimensional piecewise-linear mapping may be a useful model to study the oscillatory behaviors analytically.

The oscillatory behavior of a torus is quite analogous to Fig. 4.3 and has recently been observed in the experiments for the driven coupled p - n junction by R.V. Buskirk and C. Jeffries[15] (see Fig. 4.13). The model (4.1.1) seems to give a rather good qualitative explanation of their experiment.

We can see the process how the chaos develops to hyperchaos using a delayed logistic map. At the onset of hyperchaos (i.e., at the parameter where the second Lyapunov exponent becomes positive), however, we could not observe any singular behaviors. The "development of chaos" proceeds in the following way; appearance of chaos through a locking → the fractal dimension grows till it becomes two → width of the "belt-like" attractor increases → the unstable fixed point (from which the torus had appeared) becomes a snap-back repeller. This type of the "development of chaos" is seen in various two-dimensional mappings and is considered to be a rather universal "scenario".

Disordering property of chaos introduced in §2.8 is also seen in the delayed logistic or delayed piecewise-linear mapping. In the region $y > 0$, the mapping $(x_n, y_n) \rightarrow (x_{n+1}, y_{n+1})$ consists of rotation and stretching. Thus, the order of two orbits does not change in the region $y > 0$. If an orbit falls on the region $y < 0$, however, the folding occurs, which causes the disordering property of a chaotic orbit.

In §4.3, we considered a remarkable feature of the oscillatory instability of torus, i.e., the fractalization of torus. We observed the phenomena "torus → oscillation → fractal torus → chaos with a belt-like attractor", using a modulation map. We calculated the fractal dimension at the onset of chaos, on the basis of the method of a functional mapping. The fractal dimension seems to depend only on the rotation number

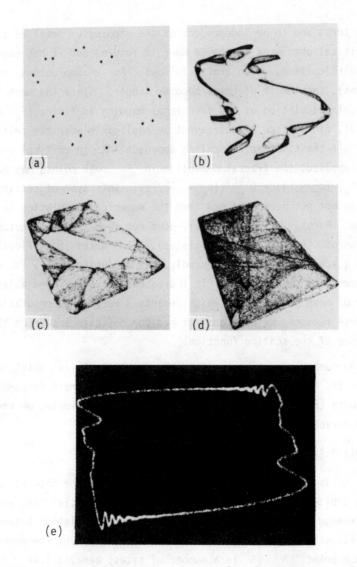

Fig. 4.13 Poincaré sections I_2 vs V_1 for the attractors of
two resistively coupled p - n junctions, where
I_j and V_j are a current and a voltage for j-th
junction (j = 1,2). The bifurcation parameter is
a drive voltage V_{os}. (a) V_{os} (V rms) = 4.409,
(b) = 4.882, (c) = 5.132, (d) = 8.958, and
(e) = 1.976 (near period 4 locking).
(cited from Ref. 15).

of the torus and to be independent of the mappings. Detailed study on the critical phenomena has to be done in future, but it has some technical difficulties, which are as follows: As ε approaches ε_c (onset of chaos), the length of torus becomes longer. Since the mesh for the numerical calculation of the functional mapping is finite (F_k or G_k in §4.3), finite size effect cannot be negligible near the critical point. If the finite-size scaling approach[8],[9] in critical phenomena can be applied, the fractal dimension in F_k mesh points may be written as $D(F_k, (\varepsilon_c - \varepsilon)) = D^{(sc)}((\varepsilon_c - \varepsilon)^a/F_k)$. This type of scaling function has not yet been obtained from the numerical simulation in our problem. First, $D(F_k, (\varepsilon_c - \varepsilon))$ does not seem to be a monotonic function of $(\varepsilon_c - \varepsilon)$ (it oscillates), which makes it difficult to obtain a scaling function. Secondly, the dimension $D(F_k, (\varepsilon_c - \varepsilon))$ decreases very slowly as F_k is increased. Thus, we need a large number of mesh points F_k, which demands a very long computation time. For these reasons, we have not yet had the definite answer on the existence of the scaling function.

The number of extremum points N_k at the critical point ε_c behaves as $N_k \sim (0.56 \pm 0.01) \times F_k$. It is of interest to compare this value with the one for the random curves. As an example, we consider a random curve generated by

$$g(i + 1) = ag(i) + n_i \quad , \tag{4.4.1}$$

where n is a white noise equally distributed in the interval $|-\sigma, \sigma|$ (σ is arbitrary). We regard the step i as a spatial site and $g(i)$ as a "random curve". If $a = 0$, there is no correlation between i-site and (i+1)-site ("white random curve"). In this curve, the number of extremum points N_ℓ (ℓ is a number of sites) behaves like

$$N_\ell = \frac{2\ell}{3} \quad , \tag{4.4.2}$$

as can be easily shown. On the other hand,

$$N_\ell = \frac{\ell}{2} \tag{4.4.3}$$

for $a = 1$ ("correlated random curve"). As 'a' is increased from 0 to 1, the coefficient (N_ℓ/ℓ) decreases from 2/3 to 1/2. Since the coefficient for our fractal curve is 0.56, it corresponds to a weakly correlated random curve $(a \sim 0.6)$. Of course, our fractal curve is not a random curve, since it is constructed by a deterministic mapping and has a self-similar structure for small scales.

The functional mapping (4.3.4) is a simple example of an infinite-dimensional mapping. Recently Yamaguti and Hata has investigated Weierstrass's function and Takagi's function using a functional equation.[10],[11] The functional equation is given by

$$g(t, \theta) = t\, g(t, \psi(\theta)) + h(\theta) \quad . \tag{4.4.4}$$

The solution for Eq. (4.4.4) is written by

$$g(t, \theta) = \sum_{n=0}^{\infty} t^n h(\psi^n(\theta)) \quad . \tag{4.4.5}$$

By choosing $h(x) = \cos \pi x$ and $\psi(x) = bx$ we obtain the Weierstrass's function, while we have Takagi's function by putting $\psi(x) = 2x$ (for $x \in [0, 1/2]$), $2 - 2x$ (for $x \in [1/2, 1]$) and $h(x) = \psi(x)/2$. Takagi's function and Weierstrass's function with the condition $0 < t < 1$, $b > 1$ and $tb \geq 1$ are nowhere differentiable and can give examples of fractal curves.

Our functional equation (4.3.2) resembles Eq. (4.4.4). In our case, however, the functional equation is nonlinear (i.e., the contraction rate t depends on $g(t, \psi(\theta))$, which makes analytical treatments difficult.

It may be also of interest to make a perturbative approach to Eq. (4.3.2) (ε is chosen as a small parameter). It is expected that the perturbation expansion loses its convergence at $\varepsilon = \varepsilon_c$. Renormalization group approach will be necessary at $\varepsilon \simeq \varepsilon_c$, which remains as a future problem.

The functional mapping (4.3.3) loses its original meaning for $\varepsilon > \varepsilon_c$, where the invariant curve does not exist. It will be of interest, however, to study the mapping for $\varepsilon > \varepsilon_c$, as a problem of a high dimensional mapping. The mapping (4.3.4) does not seem to have stable cycles with the period less than 1000, for $\varepsilon > \varepsilon_c$. The function

130

$g_n(i/F_k)$ seems to change chaotically in space (i.e., site i) and in time (i.e., step n). Thus, the map may serve as a simple turbulence model, though detailed works are left to the future (see also Chap. 7).

Oscillatory behavior of torus is also seen in the doubling of torus[5] (see Chap. 5) and in the modulated circle map[6],[12] (see Chap. 6). As for the doubling of torus, the torus seems to be fractal at the onset of chaos, which causes the stop of the doubling cascade.

Fractal torus has recently been investigated as a basin boundary of the attraction for a complex mapping.[13],[14] In our case it appears as an attractor and will be more relevant to physical observations. It will be of importance to search for this phenomenon in differential equation systems and also in experiments. A system with incommensurate modulation or a system with quasiperiodic external perturbation will be a good candidate to observe this phenomenon.

REFERENCES

*) The main contents of this chapter are published in K. Kaneko, Prog. Theor. Phys. 72 (1984) 202 and ibid 71 (1983) 1112.

1. P. Bergé, Physica Scripta T1 (1982) 71.

2. M. Sano and Y. Sawada, to appear in Turbulence and Chaotic Phenomena in Fluids, ed. T. Tatsumi (North Holland, 1984).

3. F.R. Marotto, J. Math. Anal. Appl. 63 (1978) 199.

4. Y. Takahashi, private communication.

5. K. Kaneko, Prog. Theor. Phys. 69 (1983) 1806, and 72 (1984) 202.

6. K. Kaneko, Prog. Theor. Phys. 71 (1983) 282.

7. B.B. Mandelbrot, The Fractal Geometry of Nature, (Freeman, San Francisco, 1982).

8. M.E. Fisher and N.N. Barber, Phys. Rev. Lett. 28 (1972) 1516.

9. M. Suzuki, Prog. Theor. Phys. 58 (1977) 1142.

10. M. Yamaguti and M. Hata, Hokkaido Mathematical Journal 12 (1983) 333.

11. M. Hata and M. Yamaguti, to appear in Japan J. of Appl. Math. 1.

12. See also J.P. Sethna and E.D. Siggia Physica 11D (1984) 193, where the oscillation of torus in a modulated circle map is treated. See for details §6.4.

13. S. Manton and M. Nauenberg, Comm. Math. Phys. 89 (1983) 555.

14. M. Widom, Comm. Math. Phys. 92 (1983) 121.

15. R.V. Buskirk and C. Jeffries, Phys. Rev. 31A (1985) 3332.

16. K. Ikeda, H. Daido, and O. Akimoto, Phys. Rev. Lett. 45 (1980) 709.

17. J.D. Farmer, Physica 4D (1982) 366.

18. M.C. Mackey and L. Glass, Science 197 (1977) 287.

19. R. May, Annals of NYAS 357 (1980) 267.

Appendix

In this appendix, a feature and the origin of a delayed map is given. We consider a delay differential equation[16)-19)]

$$a^{-1} \dot{x}(t) = f(x(t - t_R)) - x(t) \quad , \tag{A.1}$$

which is investigated in nonlinear optics,[16)] physiology,[18)] and ecology etc. The K-point discretization of Eq. (A.1) gives a K-dimensional mapping,

$$x_{n+1} = a f(x_{n-K+1}) + (1 - a)x_n \quad . \tag{A.2}$$

As a special case we consider the case $K = 2$. The Jacobian of the map is given by $\begin{pmatrix} 1-a & af'(x) \\ 1 & 0 \end{pmatrix}$, the eigenvalues of which are $\{(1 - a) \pm \sqrt{(1 - a)^2 + 4af'(x)}\}/2$. Thus, the fixed point $x^* = f(x^*)$ loses its stability via a Hopf bifurcation as $f'(x^*)$ becomes less than $-1/a$ and a torus appears.

If we choose $f(x) = \lambda x(1 - x)$ or $1 - Ax^2$, a delayed logistic map is obtained, which is used in Chaps. 4 and 6. If the three- or four-point discretization is used, the map which is used in Chap. 5 is obtained. After the Hopf bifurcation, collapse of tori occurs as the nonlinearity is increased in the delayed logistic maps.

Chapter 5

DOUBLING OF TORUS

Bond of Union
by *M. C. Escher*

§5.1 Discovery

In the two-dimensional mapping in Chap. 4, the doubling along the amplitude direction occurs only via a locking into a 4-cycle. In higher-dimensional mappings (or in the flow system with dimensions higher than three), the doubling of torus itself is also possible. The doubling of torus was found independently by A. Arnéodo et. al.,[1] by V. Franceschini,[2] and by the author. In this section the discovery of the doubling of torus in 3- or 4-dimensional mappings is reported.

Before proceeding to show some specific examples, we note that two types of the doubling of torus are possible for mappings, i.e., the case in which the cross section of a torus is separated (see Fig. 5.1a)) (type a) and the case in which it is still connected but two-fold (see Fig. 5.1b)) (type b).

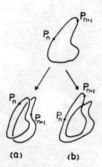

(a)   (b)

Fig. 5.1 Schematic illustration of two types of doubling of torus in k-dimensional mappings $(k \geq 3) P_n \rightarrow P_{n+1}$.

The maps investigated here are 3-point or 4-point delayed logistic map and some modified ones, which are written

(I) $X_{n+1} = AX_n + (1 - A)(1 - DY_n^2)$,

   $Y_{n+1} = Z_n$, $Z_{n+1} = X_n$; $A = 0.4$

(II) $X_{n+1} = AX_n + (1 - A)(1 - DY_n^2)$, $Y_{n+1} = Z_n$

   $Z_{n+1} = W_n$, $W_{n+1} = X_n$; $(A = 0.3$ or $A = 0.4)$

(III)
$$X_{n+1} = AX_n + (1 - A)(1 - DY_n^2),$$
$$Y_{n+1} = AY_n + (1 - A)(1 - DZ_n^2),$$
$$Z_{n+1} = AZ_n + (1 - A)(1 - DX_n^2); \quad A = 0.4,$$

and

(IV)
$$X_{n+1} = AX_n + (1 - A)(1 - DY_n^2), \quad Y_{n+1} = Z_n$$
$$Z_{n+1} = AZ_n + (1 - A)(1 - DW_n^2), \quad W_{n+1} = X_n; \quad A = 0.3,$$

where D is changed as a bifurcation parameter. (See Appendix in Chap. 4 for the meaning of delayed maps.)

These maps show the transition "fixed point → (Hopf bifurcation) → torus → doubling of torus → chaos" accompanied by frequency lockings. The doubling is of type (a) for map (I), while it is of type (b) for maps (II)~(IV).

Some examples of the attractors are given in Figs. 5.2a)~c) (for map (I)) and in Figs. 5.3a)~c) (for map (IV)), where projections onto (X,Y)-plane are depicted. Lyapunov exponents are calculated to confirm the successive bifurcations. As an example, the first and second Lyapunov exponents for the map (IV) are shown in Fig. 5.4, (the third and fourth exponents do not change drastically and they are omitted in the figure).

The parameters at which the doubling occurs are summarized in Table I. As is shown in this table, the doubling cascade stops after a finite number of times before the chaos appears (e.g., torus → 2 ⊗ torus → 4 ⊗ torus → chaos etc.). The interruption of doubling cascades of tori will be investigated in the following two sections.

§5.2 Doubling Stops by a Finite Number of Times

All examples of the doubling of torus, so far observed, show only a finite number of doublings. Thus, there arises a problem whether the doubling cascade of torus can continue infinitely in generic or not. Following essentially the idea in the paper by Ruelle and Takens,[3] we simplify and restate this problem as follows: Is the direct product state of the torus map (e.g., $Y_{n+1} = Y_n + C$ (mod 1)) and the map which

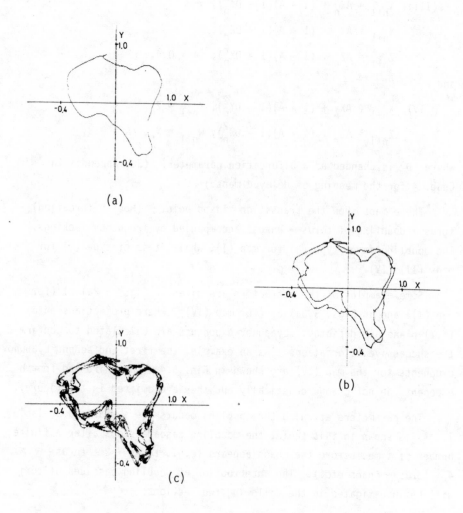

Fig. 5.2 Projection onto (X,Y)-plane of the attractor of
map (I) with A = 0.4. a) D = 2.11 (torus), b)
D = 2.16 (2 ⊗ torus), and c) D = 2.19 (chaos).
If we take only the points (X_{2n}, Y_{2n}) in Fig.
5.2b), only one torus remains. Thus, this
doubling is of type (a).

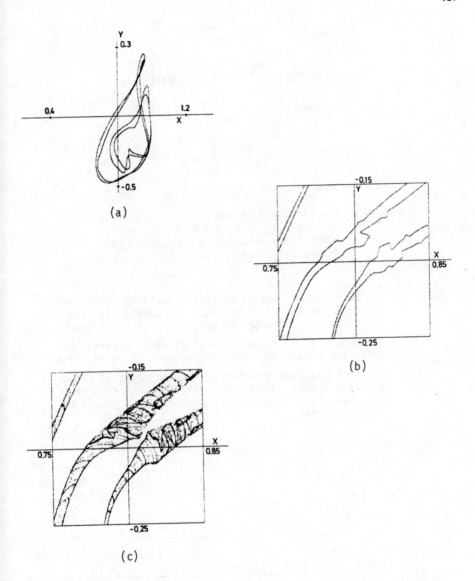

Fig. 5.3 Projection onto (X,Y)-plane of the attractor of map (IV) with A = 0.3. a) D = 1.515 (4 × torus), b) D = 1.5206 (8 × torus), and c) D = 1.5212 (chaos). For b) and c), only one part of the attractors (0.75 ≤ X ≤ 0.85 and -0.25 ≤ Y ≤ -0.15) is depicted.

138

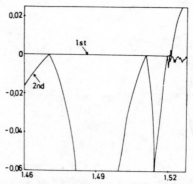

Fig. 5.4 The first and second Lyapunov exponents
for map (IV) with A = 0.3. We made
50000 iterations for calculations, with
double precision.

Table I. The parameter values at which the doubling occurs and
the value of the onset of chaos for models (I)~(IV).
The values D_k denote the parameter values at which
the doubling $2^{k-1} \times$ torus $\rightarrow 2^k \times$ torus occurs,
while the values D_c denote the onset of chaos. The
times of doubling observed before the onset of chaos
are also shown. These values are obtained from the
calculations of Lyapunov exponents and the graphs of
the attractors.

model	D_1	D_2	D_3	D_c	times of doubling
(I) A = 0.4	2.151			2.163	1
(II) A = 0.3	1.539			1.62	1
(II) A = 0.4	1.694	1.90409		1.90455	2
(III) A = 0.4	1.740			1.941	1
(IV) A = 0.3	1.470	1.5106	1.5199	1.5209	3

shows a period-doubling cascade to chaos (e.g., $X_{n+1} = 1 - AX_n^2$) stable against a structural perturbation or not? Thus, we use the method of the coupled map, which was introduced in Chap. 3 as a coupled logistic map and will also be used in Chap. 6. We performed, on this purpose, a numerical simulation of the map

$$
\begin{cases}
X_{n+1} = 1 - AX_n^2 + \varepsilon g(X_n, Y_n) & (5.2.1a) \\
Y_{n+1} = Y_n + C + \varepsilon h(X_n, Y_n) \pmod 1 , & (5.2.1b)
\end{cases}
$$

where g and h are structural perturbations, which are periodic functions about Y with period 1. We studied the following two cases in detail

$$
(I) \quad g(X_n, Y_n) = \sin(2\pi Y_n) \quad \text{and} \quad h(X_n, Y_n) = X_n \tag{5.2.2}
$$

$$
(II) \quad g(X_n, Y_n) = \sin(2\pi Y_n) \quad \text{and} \quad h(X_n, Y_n) \equiv 0 , \tag{5.2.3}
$$

where the "rotation number" C is fixed at an irrational number, e.g., at $(\sqrt{5} - 1)/2$, i.e., the inverse of the golden mean. The model (II) corresponds to the logistic map with incommensurate modulation.[4] When $\varepsilon = 0$, variables X and Y are decoupled and the transition "torus $\to$ 2 $\otimes$ torus $\to$ 4 $\otimes$ torus $\to$ 8 $\otimes$ torus $\to \cdots \to$ chaos" proceeds as A is increased.[5]

For finite ε (we have performed simulations up to $\varepsilon = 10^{-7}$ for models (I) and (II)), however, it has been found that the doubling cascade stops after a finite number of times and chaos appears from a $2^\ell \otimes$ torus (ℓ is a finite integer). The phase diagram for the map (I) is given in Fig. 5.5, which was obtained from the calculations of two Lyapunov exponents and from the patterns of the attractors. The phase diagram for the map (II) is almost similar with Fig. 5.5. As is seen from this phase diagram, the number of doublings decreases as ε is increased. The scaling relation between the number of doublings ℓ and the strength of coupling ε is shown in Fig. 5.6, where $\varepsilon_t(\ell)$ is the value at which the "tongue" in the phase diagram (see Fig. 5.5) appears for the corresponding $2^\ell \otimes$ torus-state. The scaling relation is roughly given by $2^\ell \propto \varepsilon_t^{-\chi}$ ($\chi \cong 1/3$), though, it seems to be impossible to

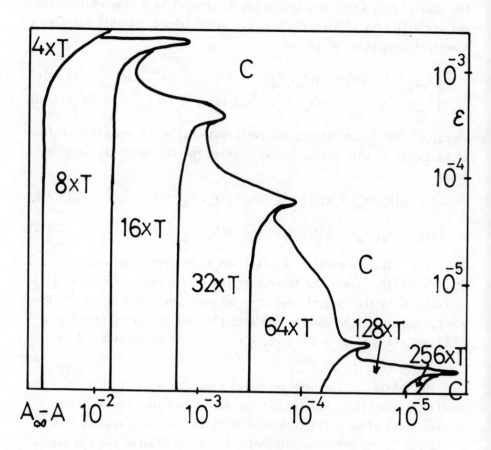

Fig. 5.5 Phase diagram for the model (I). The transverse axis
denotes $(A_\infty - A)$, where $A_\infty(= 1.401151\cdots)$ is the
value of the onset of chaos for $\varepsilon = 0$. The longitudinal axis denotes ε. "C" and "n × T" represent
chaos and $n \otimes$ torus respectively.

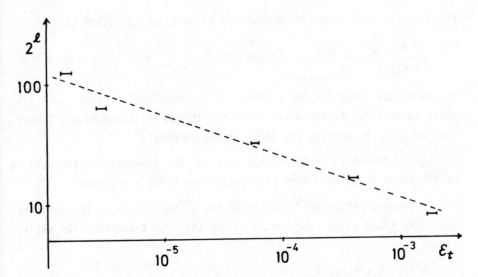

Fig. 5.6 Scaling relation between the strength of coupling
ε and the number of doublings ℓ. (See text for
the definition of $\varepsilon_t(\ell)$.)

obtain a quantitative result from this figure. Figure 5.5 gives a
qualitative explanation to the observation in §5.1 that the doubling
of torus stops after a finite number of times.

The above result shows that Feigenbaum's fixed point function[5]
is unstable against an incommensurate modulation $\varepsilon\sin(2\pi(Y_0 + nC))$.
If the incommensurate modulation is a relevant perturbation in the renormalization group framework and the eigenvalue of the renormalization
group transformation for this perturbation is κ, the scaling relation

$$X = \frac{\ln \varepsilon_t}{\ln(2^\ell)} = \frac{\ln 2}{\ln \kappa} \tag{5.2.4}$$

is derived. Renormalization group theories for noisy period-doubling
bifurcations were constructed by J.P. Crutchfield et al.[6] and B.
Schraiman et al.,[7] which show that the relevant eigenvalue for the
noisy perturbation is given by $\kappa_{noise} = 1.88995\cdots$. Thus, the scaling

relation for the noisy period-doubling bifurcations is given by

$$\frac{\ln \varepsilon_{noise}}{\ln(2^{\ell})} = \frac{\ln 2}{\ln \kappa_{noise}} = 0.366754... \quad . \tag{5.2.5}$$

The numerical value for the exponent on the doubling of torus $(\chi \sim 1/3)$ seems to be close to the above exponent, though it is beyond our numerical accuracy to confirm the value of the exponent.

Quite recently, Arnéodo[9] has made an RG argument for the scaling of the torus doubling. Here his results are briefly reviewed.

Following Feigenbaum[5], we scale the 2^n-th iterate of the logistic map. The fixed point function $F(x)$ of the RG transformation satisfies

$$\alpha^{-1}F(\alpha x) = F(F(x)) \tag{5.2.6}$$

with the scaling factor $\alpha = 1/F(1)$. The function $F(x)$ and α were already known by the Feigenbaum's theory[5].

Now let us consider a perturbation of $F(x)$ with the form $\varepsilon\omega(x, y)e^{2\pi i y}$, where $\omega(x, y)$ is 1-periodic in y, which corresponds to $\varepsilon g(x_n, y_n)$ in (5.2.1). The function $\omega(x, y)$ can be taken $\rho(x) \sin 2\pi(y + \Phi(x))$ generally.

Defining β_C and β_C' so that the renormalization operation transforms $\omega(x, y) = \rho(x) \sin 2\pi(y + \Phi(x))$ into $\alpha^{-1}\beta_C\rho(\alpha x) \sin 2\pi(y + C + \beta_C' + \Phi(\alpha x))$, we obtain two coupled equations for $\rho(x)$ and $\Phi(x)$ from the RG Eq. (5.2.6):

$$\alpha^{-1}\beta_C\rho(\alpha x) \cos 2\pi(\Phi(\alpha x) + C + \beta_C') = \rho(x)g'(g(x)) \cos 2\pi\Phi(x)$$
$$+ \rho(g(x)) \cos 2\pi(\Phi(g(x)) + C)$$

$$\alpha^{-1}\beta_C\rho(\alpha x) \sin 2\pi(\Phi(\alpha x) + C + \beta_C') = \rho(x)g'(g(x)) \sin 2\pi\Phi(x)$$
$$+ \rho(g(x)) \sin 2\pi(\Phi(g(x)) + C) \quad .$$

$$\tag{5.2.7}$$

Using a polynomial interpolation for $\rho(x)$ and $\Phi(x)$ and the known form for $g(x)$,[5] one can numerically calculate β_C and β_C' as functions of C. Here, we note that the perturbation $\omega(x, y)$ is not an eigenvector of the renormalization operation, since $\omega(x, y)$ is transformed into

$$\alpha^{-1}\beta_C \rho(\alpha x)\{\sin 2\pi(y + C + \beta_C' + \Phi(\alpha x)) \cos 2\pi(C + \beta_C')$$
$$+ \cos 2\pi(\theta + C + \beta_C' + \Phi(\alpha x)) \sin 2\pi(C + \beta_C')\} \quad . \tag{5.2.8}$$

However, one can expect that the average dilation factor is given only by the first term of (5.2.8), considering that the second term does not contribute when averaging over successive iterations. Thus, the eigenvalue κ in (5.2.4) is given by

$$\kappa = \beta_C \cos 2\pi(C + \beta_C') \quad . \tag{5.2.9}$$

From the numerical results for β_C and β_C', the eigenvalue κ is obtained as a function of the rotation number C. When C is the inverse of the golden mean, κ agrees with κ_{noise}, which is consistent with the previous result for χ in Fig. 5.6.

Arnéodo has also performed numerical computations for the Lyapunov exponent. It has a scaling form

$$L(A, \varepsilon) = \varepsilon^\chi \mathscr{L}((A - A_c)/\varepsilon^\gamma) \tag{5.2.10}$$

where $\gamma = \chi/\tau$ with $\tau = \ln 2/\ln \lambda$ and $\lambda = 4.669...$ is Feigenbaum's constant, since for $\varepsilon = 0$ it must reduce to Feigenbaum's result. The numerical calculations for χ from (5.2.10) also agree with (5.2.9).

In the supercritical region, the band merging appears for the logistic map. For the models (I) and (II), the band merging occurs only for a finite number of times just as in the case of the doubling. Thus, the interruption of doubling of torus shows a "bifurcation gap", which was observed in noisy period-doubling bifurcations.[8] In Figs. 5.7, x_n's $(500 \leq n \leq 1000)$ are plotted as functions of the bifurcation parameter A for the model (II). The bifurcation gap is clearly seen from these figures.

144

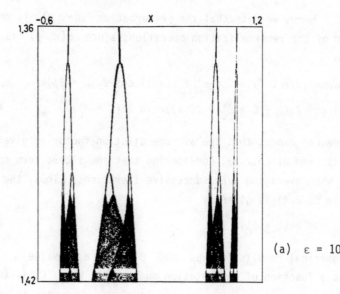

(a) $\varepsilon = 10^{-4}$

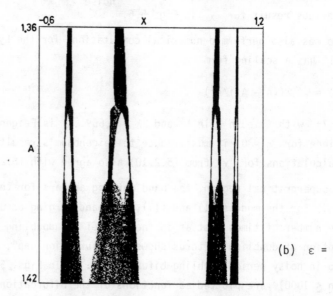

(b) $\varepsilon = 10^{-3}$

Fig. 5.7 The orbits of x_n's ($500 \leq n \leq 1000$)
for the map (5.2.1) (model (II)),
with $C = (\sqrt{5} - 1)/2$. The longitudinal axis denotes the bifurcation
parameter A.

§5.3   Mechanism of the Interruption of the Doubling Cascade

Why does the doubling cascade of tori stop by a finite number of times? What occurs at the interruption of doubling and onset of chaos? In this section, the mechanism of the interruption of the doubling cascade is investigated from two points of view, using the map (5.2.1).

The figures of the attractors are given in Figs. 5.8. As can be seen from these figures, oscillation of torus is enhanced near the onset of chaos. The torus seems to be fractal at the onset of chaos. Thus, the "fractalization of torus" will be a cause of the interruption of the doubling and the emergence of the chaos, though the detailed study has to be performed in future.

We consider the interruption of the doubling from another point of view. We consider the modulation map (II) and use a continued fraction approximation, i.e., replace C by $C_n = F_{n-1}/F_n$ (for example, F_n is a Fibonacci sequence for $C = (\sqrt{5} - 1)/2$). Then the map (5.2.1) reduces to a one-dimensional mapping

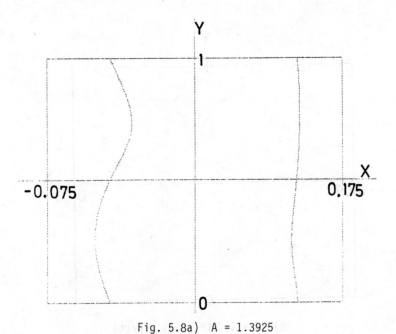

Fig. 5.8a)   A = 1.3925

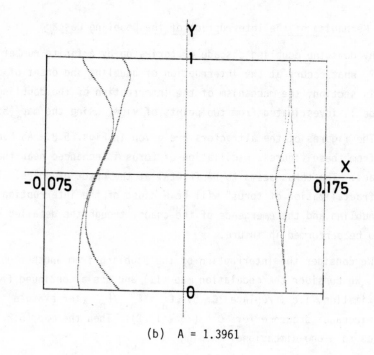

(b) A = 1.3961

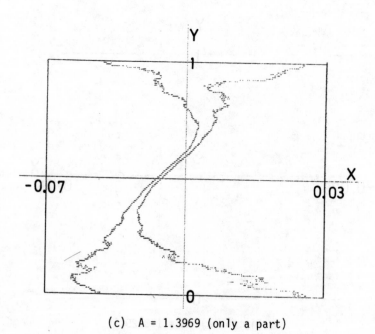

(c) A = 1.3969 (only a part)

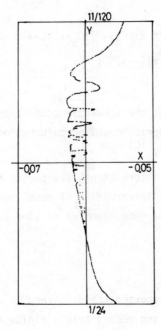

(d) A = 1.3970

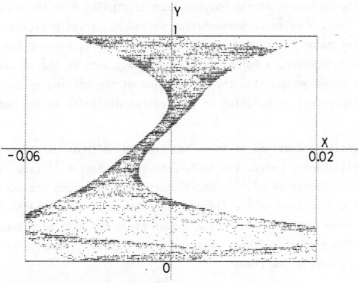

(e) A = 1.3980 (only a part)

Fig. 5.8 Attractor of the map (5.2.1) (model (II)),
with C = (√5 - 1)/2 and ε = 0.001.

$$x' = G(x)$$
$$= \varepsilon \sin(2\pi C_n(F_n - 1)) + f(\varepsilon \sin 2\pi(F_n - 2)C_n + f(\cdots$$
$$\cdots + f(\varepsilon \sin 2\pi C_n + f(x))\cdots) \quad ;$$
$$f(x) = 1 - Ax^2 \qquad\qquad\qquad (5.3.1)$$

for $Y_0 = 0$. The map (5.3.1) satisfies the Schwarzian condition, but it is not a unimodal mapping. Thus, period-doubling bifurcations in the map (5.3.1) may not continue infinitely.[10]

As a matter of fact, we observed intermittent-like bursts between the valleys of $G(X)$, which cause the interruption of doublings and the transition to chaos. (See Fig. 5.9 for some examples of time series by the map $G(X)$).

§5.4 Discussion

In the present chapter, we have reported the discovery of doubling of torus and have shown that the doubling occurs only a finite number of times. The mechanism of the interruption is studied from three points of view, i.e., a relevant perturbation in an RG transformation, fractalization of torus, and intermittent-like bursts between two valleys. There remain however, a lot of future problems, such as the relationship among the three viewpoints, the computation of the scaling exponent by RG, and numerical calculation of the fractal dimension by the method in §4.3.

Recently, a doubling of torus has been found in an experiment of the Rayleigh-Bénard convection by M. Sano and Y. Sawada[11] (see Fig. 5.10) and by Haucke et al.[12] Oscillatory behavior along the amplitude direction can be seen in the figure. The number of the doubling observed in their experiment was one. Doubling of torus has also been found in a simulation of Navier-Stokes equation by H. Yahata,[13] where the doubling occurs only a few times.

In the present chapter a method of coupled maps is used to study the stability of a direct product state. The method will be useful to investigate the stability of various direct product states and of some

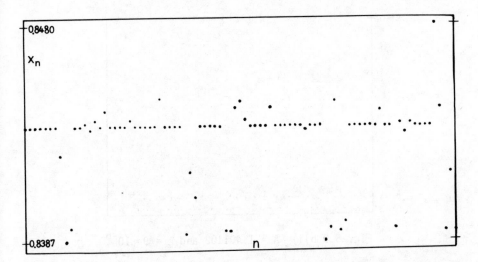

Fig. 5.9 a)i) A = 1.4006 and $\varepsilon = 2 \times 10^{-4}$

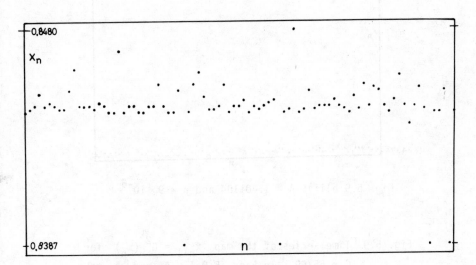

Fig. 5.9 a)ii) A = 1.40065 and $\varepsilon = 2 \times 10^{-4}$

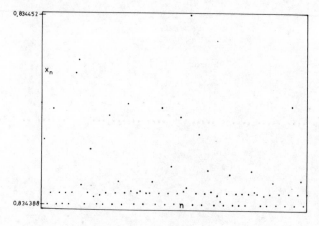

Fig. 5.9 b)i) A = 1.401102 and $\varepsilon = 9 \times 10^{-6}$

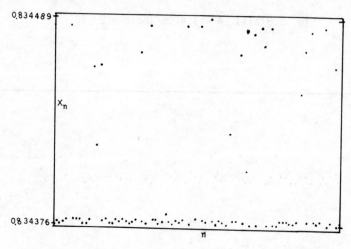

Fig. 5.9 b)ii) A = 1.401104 and $\varepsilon = 9 \times 10^{-6}$

Fig. 5.9 Time series of the map $x_{n+1} = G^{2^k}(x_n)$ for
C = 55/89. In Figs. 5.9a), A and ε take
the values at which 32 $\otimes$ torus collapses
(k = 5), while they take the values at which
64 $\otimes$ torus collapses (k = 6) in Figs. 5.9b).
(G(x) is given in Eq. 5.3.1)

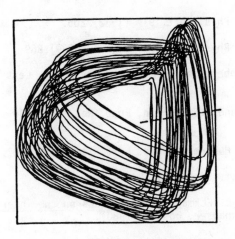

Fig. 5.10 Phase portrait and Poincaré section for the
doubling of torus. The aspect ratio is 3.5
and the Prandtl number is 5.5. Complicated
twisting of closed curves is caused by the
projection onto 2-dimensional plane. (cited
from Ref. 11)

bifurcation sequences. If we replace Eq. (5.2.1a) by $Y_{n+1} = 2Y_n \pmod 1$, for example, we can treat the "doubling cascade of chaos to hyperchaos" and can show that the doubling of chaos occurs only a finite number of times. On the other hand, we can treat the torus intermittency by replacing Eq. (5.2.1b) by the map which gives the intermittent transition to chaos.[14] This study has been performed by H. Daido[15] quite recently. The coupled maps will remain to be effective measures to study the stability of various direct product states.

REFERENCES

*) The contents of this chapter are mainly based on K. Kaneko, Prog. Theor. Phys. 69 (1983) 1806 and Prog. Theor. Phys. 72 (1984) 202.

1. A. Arnéodo, P.H. Coullet and E.A. Spiegel, Phys. Lett. 94A (1983) 1.

2. V. Franceschini, Physica 6D (1983) 285.

3. D. Ruelle and F. Takens, Comm. Math. Phys. 20 (1971) 167.

4. Logistic map with commensurate parametric modulation was studied

152

by M. Lücke and Y. Saito, Phys. Lett. 91A (1982) 205.

5. M.J. Feigenbaum, J. Stat. Phys. 19 (1978) 25, 21 (1979) 669.

6. J.P. Crutchfield, M. Nauenberg and J. Rudnick, Phys. Rev. Lett. 46 (1981) 933.

7. B. Schraiman, C.E. Wayne and P.C. Martin, Phys. Rev. Lett. 46 (1981) 935.

8. J.P. Crutchfield and B.A. Huberman, Phys. Lett. 74A (1980) 407.

9. A. Arnéodo, Phys. Rev. Lett. 53 (1984) 1240.

10. See e.g., P. Collet and J.P. Eckman, Iterated Maps on the Interval as Dynamical Systems, Birkhäuser, Boston (1980).

11. M. Sano and Y. Sawada, Chaos and Statistical Methods ed. Y. Kuramoto, Springer (1983).

12. H. Haucke, Y. Maeno and J.C. Wheatley in Proc. of 7th Int. Conf. on Low Temperature Physics, North-Holland (1984).

13. H. Yahata, private communication.

14. Y. Pomeau and P. Manneville, Comm. Math. Phys. 74 (1980) 189.

15. H. Daido, Prog. Theor. Phys. 71 (1984) 402.

Chapter 6

FATES OF THREE-TORUS

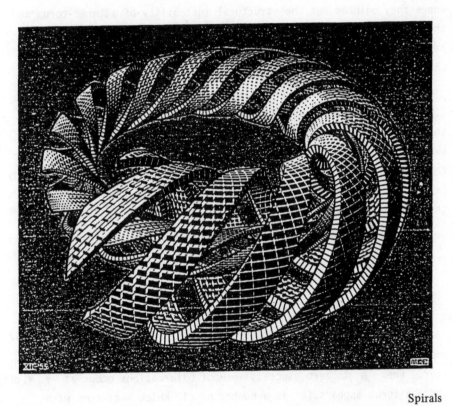

Spirals
by *M. C. Escher*

154

§6.1 Introduction

Quasiperiodic motion with three incommensurate frequencies (which we call "three-torus" in the present book) is considered to appear via a Hopf bifurcation of a torus (the term "torus" is used only for a two-torus). In 1971, however, Ruelle and Takens wrote an important paper,[1] where they pointed out the structural instability of a three-torus and the emergence of a strange attractor. It denied the picture of turbulence by Landau and opened a new era for the study of low-dimensional chaos.

In experiments, however, three-tori have seemed to be observed, by J.P. Gollub and S.V. Benson,[2] A. Libchaber et al.,[3],[4] and so on,[5] though, it is not so easy to distinguish three-tori from (two)-tori experimentally. Numerical studies on three-tori, are very few except the simulation of a 56-mode truncation of the Navier-Stokes equation by Yahata.[6] Thus, it will be of importance to study the features of a three-torus in a simple system.

The successive Hopf bifurcations can be modeled by the equations

$$\dot{w}_j = (\gamma_0 - d(j))w_j - (g_1 + ig_2)|w_j|^2 w_j$$
$$+ \varepsilon h_j(w_1, w_1^*, w_2, w_2^*, \cdots, w_N, w_N^*) \quad (j = 1,2, \cdots, N)$$

$$(6.1.1)$$

where γ_0 is a bifurcation parameter and w is a complex order parameter. When ε is zero, successive Hopf bifurcations occur at $\gamma_0 = d(i)$ and a k-torus appears (k is a number of i which satisfies with $\gamma_0 > d(i)$). As the perturbation εh sets in, the direct product state may become unstable, and the k-torus can be destructed. Though it will be of interest to study Eqs. (6.1.1), we make a further simplification in the present chapter, that is, we make use of a coupled map.

Since the dimension is reduced by one by taking a Poincaré map from a flow system, the three-torus in a flow corresponds to an attractor in a map with the first and second Lyapunov exponents vanishing. Thus, we choose the following mapping as a model of a three-torus;

$$x_{n+1} = f(x_n, y_n; a) + \varepsilon h_1(x_n, y_n, z_n, w_n)$$

$$y_{n+1} = g(x_n, y_n; a) + \varepsilon h_2(x_n, y_n, z_n, w_n)$$

$$z_{n+1} = f(z_n, w_n; a') + \varepsilon h_3(x_n, y_n, z_n, w_n)$$

$$w_{n+1} = g(z_n, w_n; a') + \varepsilon h_4(x_n, y_n, z_n, w_n) \qquad (6.1.2)$$

where the mappings $x_{n+1} = f(x_n, y_n; a)$, $y_{n+1} = g(x_n, y_n; a)$ show the transition from torus to chaos as the bifurcation parameter a is increased, and εh_i ($i = 1,2,3,4$) are perturbations. In §2, we take the delayed logistic model (see Appendix in Chap. 4) for $\{f, g\}$, to study the stability of a three-torus.

If our interest is restricted only to the phase motion of the three-torus, a further simplification may be possible just as in the study of (two)-torus by the map (2.1.2). Thus, a coupled circle map

$$\theta_{n+1} = \theta_n + A \sin (2\pi\theta_n) + D + \varepsilon \sin (2\pi\varphi_n)$$

$$\varphi_{n+1} = \varphi_n + B \sin (2\pi\varphi_n) + C + \varepsilon' \sin (2\pi\theta_n) \qquad (6.1.3)$$

will be of relevance to the study of the phase motion of a three-torus. In §3, the simplest case among the mappings (6.1.3) is investigated, that is the "modulated circle map". Various lockings into tori are found. The rotation number as a function of a bifurcation parameter forms a "double devil's staircase", which will also be shown in §3. In §4, more general cases of the coupled circle maps will be treated in detail, where the competition among T^3, T^2 chaos and lockings will be investigated.

The main purpose of the present chapter is to give only qualitative features of a three-torus with various figures. In this sense, the understanding of three-torus remains rather incomplete and many problems are left to the future. Discussions including these problems will be given in §5.

§6.2 Three-torus in a Four-dimensional Mapping

In this section, we show numerical results on the 'coupled delayed logistic map', to discuss the stability of a three-torus. Delayed-logistic map is given by

$$x_{n+1} = Ax_n + Dx_{n-1}(1 - x_{n-1}) \tag{6.2.1}$$

which shows a transition "fixed point → (Hopf bifurcation) → torus → locking → chaos → hyperchaos" as D is increased (see Chap. 4). According to the idea in §1, we couple two delayed-logistic mappings and add a coupling term. The model equations constructed in this way, are given by

$$x_{n+1} = Ax_n + D_1 x_{n-1}(1 - x_{n-1}) + \varepsilon h_1(x_n, x_{n-1}, z_n, z_{n-1})$$
$$z_{n+1} = Az_n + D_1 z_{n-1}(1 - z_{n-1}) + \varepsilon h_2(x_n, x_{n-1}, z_n, z_{n-1}) \tag{6.2.2}$$

Eq. (6.2.2) is a 4-dimensional map $(x_n, y_n, z_n, w_n) \to (x_{n+1}, y_{n+1}, z_{n+1}, w_{n+1})$, where $y_n \equiv x_{n-1}$ and $w_n \equiv z_{n-1}$. In the present chapter the value A is fixed at 0.4, though the qualitative behaviors are insensitive to the change of A. As is given in the Appendix, the delayed logistic map shows a Hopf bifurcation at $D = D_c = 3 - 2A = 2.2$ and a torus appears for $D > D_c$, which is destroyed via lockings and chaos appears for $D \gtrsim 2.59$. Thus, the direct product state, such as $T \otimes P$, $T \otimes T$, $C \otimes T$, $C \otimes C$ (P, T, and C denote periodic state (cycle), torus, and chaos respectively) exists at the corresponding values of D_1 and D_2, for $\varepsilon = 0$.

As the couplings set in, the direct product state can become unstable. Here, we chose the perturbations as $h_1 = z_n - z_{n-1}$ and $h_2 = x_{n-1} - x_n$ and made simulations of the map (6.2.2) for various values of D_1, D_2 and ε, to study the stability of the direct product state.

In Fig. 6.1, a rough phase diagram for the map is given, where the parameter D_1 is taken as $D_1 = D_2 + 0.1$. We classify the attractors into cycle (P), torus (T), 3-torus (3T), chaos, and hyper^k-chaos (kC),[7] by calculating the Lyapunov exponents from the first to the fourth.[8] Examples of Lyapunov exponents as a function of D_2 for $\varepsilon = 10^{-3}$, 5×10^{-3}, and 10^{-2} are given in Figs. 6.2a)-c), while examples of the attractors for $\varepsilon = 5 \times 10^{-3}$ are shown in Figs. 6.3a)-f), where the projections onto (x_n, z_n)-plane are depicted. At the rest of this

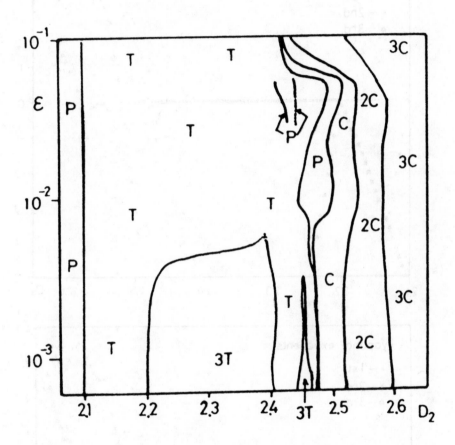

Fig. 6.1 Phase diagram of the map (6.2.2) with $D_1 = D_2 + 0.1$, $h_1 = z_n - z_{n-1}$, and $h_2 = x_{n-1} - x_n$. The notations P, T, 3T, C, and nC denote periodic state (cycle), torus, 3-torus, chaos, and hyperchaos with n positive Lyapunov exponents. This diagram was obtained from the calculation of four Lyapunov exponents.

158

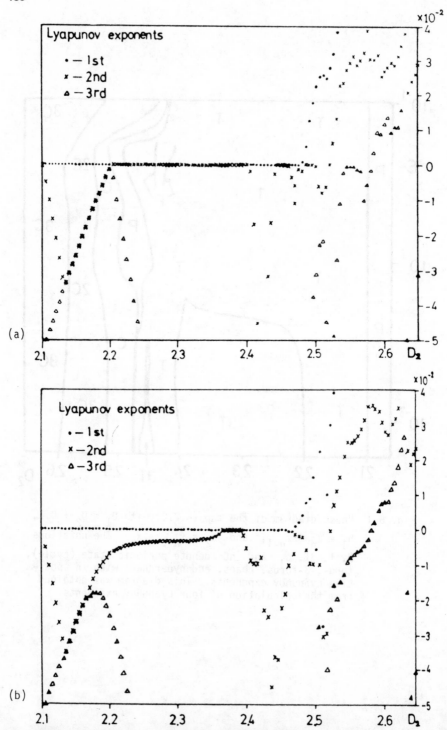

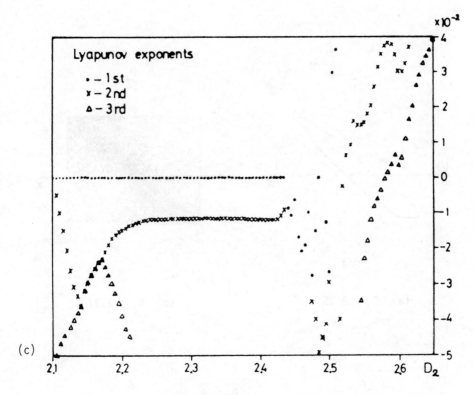

(c)

Fig. 6.2   Lyapunov exponents for the map (6.2.2)
with $D_1 = D_2 + 0.1$, $h_1 = z_n - z_{n-1}$,
and $h_2 = x_{n-1} - x_n$,   which were
calculated by the method due to Shimada
and Nagashima[8]. The values of ε are
10^{-3} (a), 5×10^{-3} (b),   and 10^{-2} (c)
respectively. The first ($\bullet$), second (X),
and third ($\triangle$) Lyapunov exponents are
plotted for $2.1 \leq D_2 \leq 2.65$,   while the
fourth exponent is large in magnitude
(negative) and is omitted.

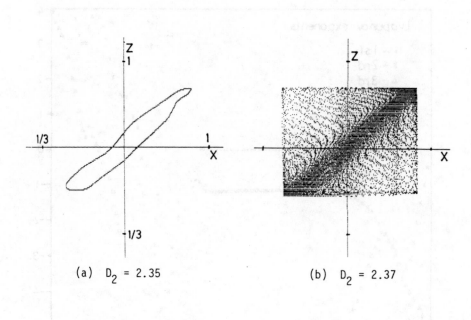

(a) $D_2 = 2.35$ (b) $D_2 = 2.37$

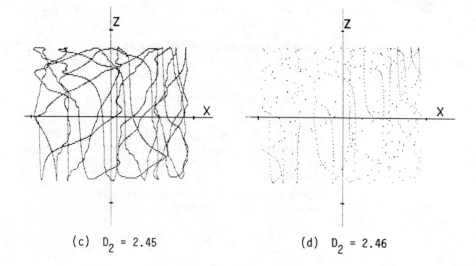

(c) $D_2 = 2.45$ (d) $D_2 = 2.46$

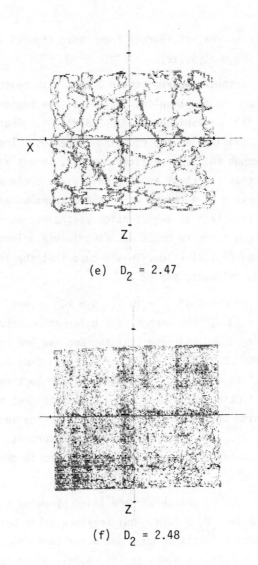

(e) $D_2 = 2.47$

(f) $D_2 = 2.48$

Fig. 6.3 Projection onto (x_n, z_n)-plane of the attractor
of the map (6.2.2) with $D_1 = D_2 + 0.1$,
$h_1 = z_n - z_{n-1}$, and $h_2 = x_{n-1} - x_n$. The
parameter ε is fixed at 5×10^{-3}.

section, we describe what is learned from these figures and other numerical results of the map (6.2.2).

As is seen in these figures, there exists a three-torus for a small coupling ε. As the coupling increases, the region of three-torus decreases till it vanishes for $\varepsilon \gtrsim 9 \times 10^{-3}$. Thus, the three-torus loses its stability as the coupling increases. The attractor that emerges through this instability, however, is not a chaos but a torus. We note that the chaos appears only via a cycle which has appeared as a locking of a torus, which again appeared as a locking of a three-torus. This fact is an extension of the recent observation that the transition from torus to chaos occurs only via a locking into cycle for the circle map (2.1.2) (i.e., in the case that the instability in phase dynamics is relevant; see §2.7).

Let us see the cases of $\varepsilon = 10^{-3}$, 5×10^{-3}, and 10^{-2} in more detail. For $\varepsilon = 10^{-3}$, the second Hopf bifurcation occurs at $D_2 = D_c = 2.2$ (see Fig. 6.2a) and note that the 2nd and 3rd Lyapunov exponents are degenerate for $2.13 \leq D_2 \leq 2.2$) and a three-torus appears. As the nonlinearity D_2 is increased, the region of the locking into torus increases. The locking into torus occurs via a tangent bifurcation. Thus, the mechanism of the locking is same as for the locking into cycle (see Chap. 2). The appearance of chaos occurs at $D_2 \cong 2.48$. The mechanism of onset of chaos will be disscussed in detail for $\varepsilon = 5 \times 10^{-3}$.

For $\varepsilon = 5 \times 10^{-3}$, the second and third Lyapunov exponents are degenerate for $2.14 \leq D_2 \leq 2.18$, but they are split before $D_2 = 2.2$, and the second Hopf bifurcation does not occur (see Fig. 6.2b)). The attractor at $D_2 = 2.35$ is given in Fig. 6.3a), which is regarded as a locking from a three-torus. This locking disappears via a tangent bifurcation and a three-torus appears (see Fig. 6.3b)). As we increase D_2 further, various lockings into tori appear (see Fig. 6.3c)). These lockings are characterized by the rotation numbers ρ_x and ρ_z, which are defined by

$$\lim_{n \to \infty} \frac{1}{2\pi n} \sum_{i=1}^{n} \arg(\overrightarrow{P_i P_{i+1}}, \overrightarrow{P_{i+1} P_{i+2}})$$

where $P_i = (x_i, x_{i-1})$ or (z_i, z_{i-1}) respectively. The locking in Fig. 6.3a) is characterized by $\rho_x = \rho_z$. The behaviors of lockings with rotation numbers are investigated in detail in §3 for a simplified model. For $D_2 \gtrsim 2.458$, the chaos appears (Fig. 6.3e)) via a locking into a cycle (Fig. 6.3d)). As D_2 is increased further, the second and third Lyapunov exponents become positive successively. We note that the picture of a "direct product state" is recovered (see e.g. Fig. 6.3f)) as the Lyapunov exponents get large.

For $\varepsilon = 10^{-2}$, the region of the locking into a torus increases, and the three-torus has disappeared (see Fig. 6.2c)). Transition from torus to chaos via lockings into cycle occurs for $D_2 \cong 2.50$, just as in the case for two-dimensional mappings.

Numerical simulations on other cases, such as $D_2 = D_1$ or $\varepsilon < 0$ or other types of couplings h_1 and h_2 were also performed. For $D_2 = D_1$, the locking into the torus with $\rho_x = \rho_z$ is more dominant and the three-torus is not observed for $\varepsilon \gtrsim 10^{-3}$. For the case with $\varepsilon < 0$ or with other types of perturbations (such as $h_1 = z_n - x_n$, $h_2 = x_n - z_n$ or $h_1 = (z_n - z_{n-1})^\alpha$ and $h_2 = (x_{n-1} - x_n)^\alpha$ ($\alpha = 1/2$ or 2)), the qualitative features are same as the above cases.

Thus, the three-torus exists for a small coupling, which becomes feasible to lock into torus as the coupling is increased and which collapses above some "critical" coupling.

§6.3 Double Devil's Staircase in Modulated Circle Map

In this section we study the phase motion of a three-torus with the use of a coupled circle map given in Eq. (6.1.3). Here, the simplest case $B = \varepsilon' = 0$ is treated, that is, the "modulated circle map",

$$\begin{cases} \theta_{n+1} = \theta_n + A \sin (2\pi\theta_n) + D + \varepsilon \sin (2\pi\varphi_n) \quad (\text{mod } 1) \\ \varphi_{n+1} = \varphi_n + C \quad (\text{mod } 1) \end{cases}$$

$$(6.3.1)$$

The parameter C is chosen to be irrational, which is fixed at $(\sqrt{5} - 1)/2$, i.e., the inverse of the golden mean.

One of the Lyapunov exponents is always zero and if the other is also zero, then the attractor is a three-torus (to be precise, it is the phase part of the Poincaré map of a three-torus, but we call it three-torus for simplicity in the present chapter).

The map (6.3.1) is invertible for $A < 1/(2\pi)$. The attractor for $A < 1/(2\pi)$ is a torus or a three-torus which is understood as follows: First, we approximate C by F_{n-1}/F_n using the continued fraction expansion (for the case $C = (\sqrt{5} - 1)/2$, F_n is a Fibonacci sequence). Iterating the map (6.3.1) F_n times, we obtain a one-dimensional map, which is an invertible circle map and its attractor is a cycle or torus. (cf. §2.1) Taking a limit $n \to \infty$, we can confirm the above statement.

The rotation number for φ is fixed at C and the rotation number ρ_θ for θ is a monotonic (in a wide sense) function of D. We conjecture that for $A < 1/2\pi$, the locking from three-torus into torus occurs if and only if ρ_θ takes a value $q/p + Cs/r$ (p, q and r, s are relatively prime integers respectively). The above conjecture is a natural extension of the theorem on the locking of the torus and seems to hold in numerical results. In our case, we note that q or s can take a negative value even if ρ_θ is positive.

Examples of the attractors are given in Figs. 6.4a)-d) ($\varepsilon = 10^{-2}$). The locking with $\rho_\theta = C$ occurs for $D \geq 0.6054$ (Fig. 6.4b)). As is seen in Fig. 6.4a), a kind of "intermittent" behavior is seen just a smaller value of D than 0.6054, which is typical for the tangent bifurcation (see also Fig. 6.3b)).

Locking with $\rho_\theta = 5/11 + 3C/11$ is given in Figs. 6.4c), while the attractor of three-torus is shown in Figs. 6.4d). The values p, q, r, s (where $\rho_\theta = q/p + Cs/r$) are given in Tables 6.Ia) ($\varepsilon = 10^{-2}$), 6.Ib) and 6.Ic) ($\varepsilon = 10^{-1}$). We note that the various lockings with

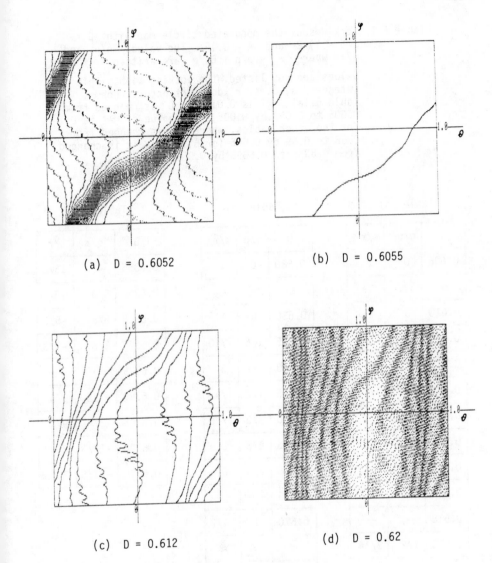

(a) D = 0.6052

(b) D = 0.6055

(c) D = 0.612

(d) D = 0.62

Fig. 6.4 Attractor of the modulated circle map (6.3.1)
with C = (√5 - 1)/2, A = 0.15 and ε = 0.01.
The points (θ_i, φ_i) (50000 ≤ i ≤ 25000) with
the initial values (θ_0, φ_0) = (0.5, 0.5) are
plotted.

Table 6.I Lockings of the modulated circle map with $C = (\sqrt{5} - 1)/2$ and $A = 0.15$. The values q/p and s/r where $\rho_\theta = q/p + Cs/r$ are written. If the values are not listed, the locking with simple integers q, p, s, r does not occur. For Table 6.Ia), ε is 0.01 and D is changed from 0.606 to 0.690 by 0.005. The value of ε is 0.1 for the other tables, where D is changed from 0.58 to 0.69 by 0.005 for 6.Ib) and it is changed from 0.6355 to 0.6395 by 0.0005 for 6.Ic).

Table a)

D	q/p	s/r
0.606	0	1
≈	≈	≈
0.610	0	1
0.611	5/18	-5/9
0.612	5/11	3/11
0.613	5/9	1/11
0.614	5/7	1/7
0.615	5/6	-1/3
0.616	5/6	-1/3
0.617	5/6	-1/3
0.618	——	——
0.619	1/2	3/14

Table b)

D	q/p	s/r
0.580	0	1
≈	≈	≈
0.630	0	1
0.635	1/5	7/10
0.640	5/2	-3
0.645	1/2	1/4
0.650	3/5	1/10
0.655	2/3	0
0.660	2/3	0
0.665	——	——
0.670	1	-1/2
≈	≈	≈
0.685	1	-1/2
0.690	——	——

Table c)

D	q/p	s/r
0.6355	-4	15/2
0.6360	5/3	-5/3
0.6365	5/3	-5/3
0.6370	5/3	-5/3
0.6375	——	——
0.6380	-3/4	9/4
0.6385	-3/4	9/4
0.6390	-16/3	29/3
0.6395	-11/5	23/5

ρ_θ = q/p + Cs/r form rather complicated structures. The rotation number ρ_θ as a function of D is given in Figs. 6.5a) ($\varepsilon = 10^{-3}$), 6.5b) ($\varepsilon = 10^{-2}$), and 6.5c) ($\varepsilon = 10^{-1}$). As is seen in Figs. 6.5 and Tables I, the region of the lockings with ρ_θ = q/p + Cs/r (s $\neq$ 0) increases as ε is increased, while the region of the locking with s = 0 (e.g. ρ_θ = 2/3) decreases. Since the number of elements to construct the staircase in Fig. 6.5 are two (i.e., C and 1), it may be called "double devil's staircase". The second Lyapunov exponent as a function of D is given in Figs. 6.6a) and b).

As is seen in Figs. 6.5 and 6.6, the region of locking increases as the coupling ε is increased. This is typically shown in Fig. 6.7, where the ratio of the locking region is plotted as a function of log ε. The above observation agrees well with the decrease of the region of three-torus due to the coupling, found in the simulations in §6.2.

Before closing this section, we give some observations about the transition to chaos. In Figs. 6.8a) and b), the instability of torus with ρ_θ = C and the transition to chaos are depicted.

The oscillation of torus is remarkably seen in Fig. 6.8a). This type of the oscillation of torus is also observed in other two-dimensional mappings and in the doubling of torus (see Chaps. 4 and 5). The oscillation is magnified as A is increased, which causes the transition to chaos. Thus, the fractalization of torus in Chap. 4 seems to be a dominant mechanism of the onset of chaos.

§6.4   Chaos from T^3 in a Coupled Circle Map

In the present section, we consider the phase motion of a three-torus in a little more detail, using a generic coupled circle map, which is given by

$$x_{n+1} = x_n + r_x + A \sin (2\pi x_n) + B \sin (2\pi y_n)$$

$$y_{n+1} = y_n + r_y + C \sin (2\pi x_n) + D \sin (2\pi y_n) \qquad (6.4.1')$$

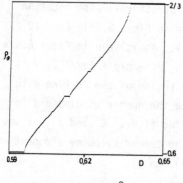

(a) $\varepsilon = 10^{-3}$

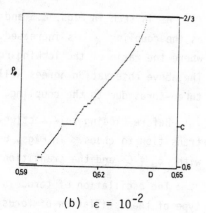

(b) $\varepsilon = 10^{-2}$

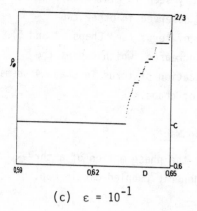

(c) $\varepsilon = 10^{-1}$

Fig. 6.5 The rotation number ρ_θ as a function of D
for the modulated circle map (6.3.1) with
$C = (\sqrt{5} - 1)/2$ and $A = 0.15$.

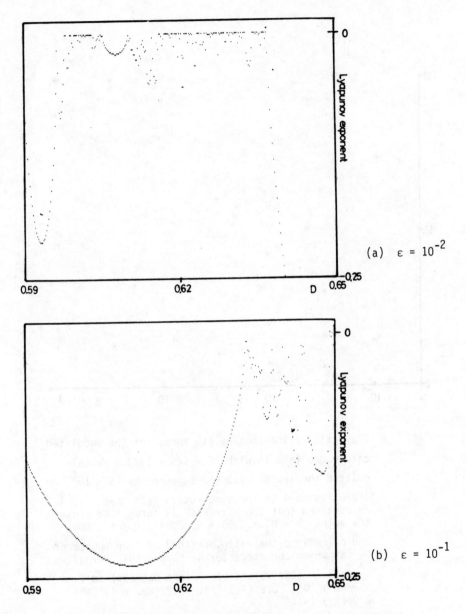

Fig. 6.6 The second Lyapunov exponent of the modulated
circle map (6.3.1) with C = ($\sqrt{5}$ - 1)/2 and
A = 0.15, while the first Lyapunov exponent
is always zero (trivial). It is plotted as
a function of D.

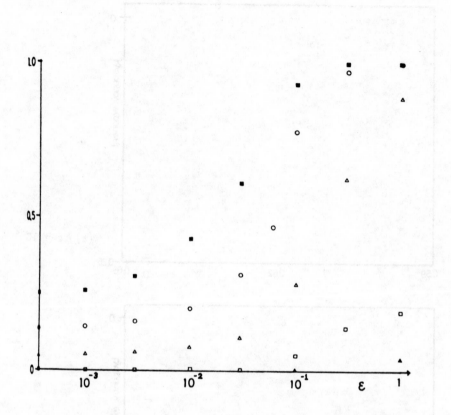

Fig. 6.7 The ratio of the locking (to torus) of the modulated circle map (6.3.1) with $C = (\sqrt{5} - 1)/2$. We calculated the second Lyapunov exponent by 5×10^4 iterations and if the exponent is less than -10^{-4}, we regarded that the attractor is torus. We chose the value $D = D_i = 0.55 + 0.0005i$ $(0 \leq i \leq 600)$ and calculated the ratio, defined by (the number of D_i at which the attractor is torus)/601, for given ε and A. The values of A are $0.9/(2\pi)$ (■), $0.8/(2\pi)$ (○), $0.6/(2\pi)$ (△), $0.2/(2\pi)$ (□), and $0.06(2\pi)$ (▲).

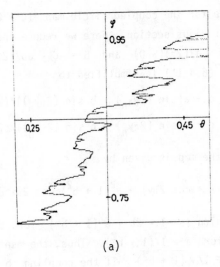

(a)

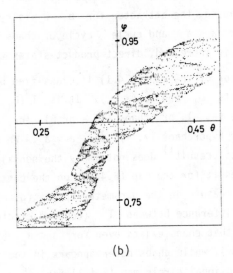

(b)

Fig. 6.8 The attractor of the modulated circle map
(6.3.1) with ε = 0.05, D = 0.615 and
C = (√5 - 1)/2. Only a part of the
attractor is shown. The values of A
are 0.178 (6.8a)) and 0.182 (6.8b)).

A nongeneric case for the coupled circle map $(C = D = 0)$ was investigated in the previous section. Here we reduce the number of parameters by taking $A = D = a/(2\pi)(> 0)$ and $B = -C = ab/(2\pi)$ and $r_x = r_y = d$. Thus, the map is (6.4.1') is simplified to

$$x_{n+1} = x_n + d + a(\sin(2\pi x_n) + b\sin(2\pi y_n))/(2\pi)$$

$$y_{n+1} = y_n + d + a(\sin(2\pi y_n) - b\sin(2\pi y_n))/(2\pi) \quad . \qquad (6.4.1)$$

The Jacobian of the map is given by

$$1 + a(\cos 2\pi x + \cos 2\pi y) + a^2(1 + b^2)\cos 2\pi x \cos 2\pi y \quad ,$$

the minimum of which is $1 - 2a + a^2(1 + b^2)$ for $a < 1/(1 + b^2)$ and $1 - a^2(1 + b^2)$ for $a > 1/(1 + b^2)$. Thus, the map (6.4.1) is invertible for $a < a_c \equiv 1/\sqrt{(1 + b^2)}$. If the coupling b vanishes, the map is reduced to two independent circle maps with identical parameters. According to Chap. 2, the attractor for $b = 0$ is cycle $\otimes$ cycle or $T^2 \otimes T^2 (T^3)$ for $a < a_c$ and cycle $\otimes$ cycle or chaos $\otimes$ chaos for $a > a_c$. As the coupling is increased, direct product states can become unstable.

The attractor of the map (6.4.1) is classified by the signs of Lyapunov exponents L_1 and $L_2(< L_1)$. It is $T^3(L_1 = 0, L_2 = 0)$, $T^2(0, -)$, cycle $(-, -)$ or chaos $(+, -$ or $0)$ for $a < a_c$ and T^2, cycle, chaos, or hyperchaos $(+, +)$ for $a > a_c$. We again note that Ruelle and Takens' result[1] does not imply the nonexistence of T^3. The numerical result for the map (6.4.1), on the contrary, shows that T^3 has a large measure in the parameter space (d, b) for small a. The remarkable difference between T^3 (coupled circle map) and T^2 (circle map) is that chaos exists even for $a < a_c$ (i.e., in the invertible regime), while chaos never appears in the invertible regime for the one-dimensional circle map (2.1.1) for T^2.

It is a little bit difficult to distinguish T^3, T^2, and chaos accurately by numerical methods. We calculated the two Lyapunov exponents L_1 and L_2 for the map with $b = 0.1$ and $b = 0.5$, by iterating the map (6.4.1) 10^5 times after dropping initial 2×10^4 times of iterations. We change the parameter d by 0.0005 for

$0 < d < 0.5$ for a given a (thus, 10^3 points of d's are chosen) and counted the number of T^3, T^2, cycle and chaos. Here, the Lyapunov exponent is regarded as zero if its magnitude is less than 10^{-4}. The ratios of the four types of attractors (i.e., the number of the parameter d at which each attractor appears, divided by 10^3) are shown in Figs. 6.9a) and b).

For small a, the magnitude of Lyapunov exponents is small in general and the distinction of attractors by the above criterion is not accurate. Especially it is difficult to distinguish T^3 from T^2 with small negative L_2. Thus, the results for $a < 0.5a_c$ might not be taken seriously. The increase of the measure of T^2 for small a might also be due to the lack of iterations.

The following aspects can be seen from Figs. 6.9.

(1) T^3 stably exists even for a system with a finite coupling and a finite nonlinearity. The ratio is rather large for a weak nonlinear system. It decreases as the increase of a and vanishes at $a \sim a_c$.

(2) Lockings to cycles increase as a approaches a_c. The increase is the most remarkable change as a function of a, which can be seen from Fig. 6.9. Though the ratio of lockings is less than unity at $a = a_c$ owing to the existence of chaos and T^2, it is the largest among the ratios for the four types of attractors. In this sense, lockings play a major role for the instability and disappearance of a three-torus.

(3) Chaos appears even for $a < a_c$. Its measure, however, is very small for small a. Chaos was not found for $a < 0.7a_c$ (for b = 0.1) or for $a < 0.6a_c$ (for b = 0.5). In other words, chaos of Ruelle and Takens' type[1] cannot be observed for $a < 0.6a_c$. The result agrees with the simulation of 4-dimensional mappings in §6.2, where chaos cannot be observed in the parameter region where T^3 is dominant. We cannot, of course, deny the mathematical existence of chaos for $a < 0.7a_c$ by numerical methods. It may be concluded, however, that chaos does not exist "physically" in the weak non-

174

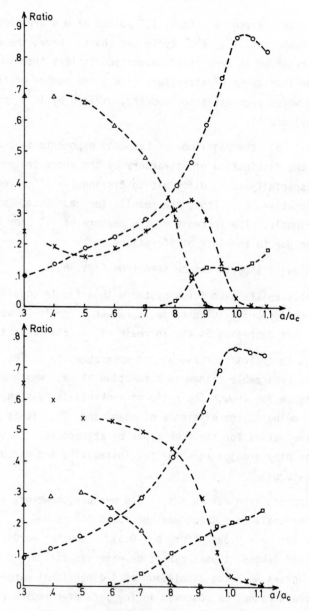

Fig. 6.9 The ratio of attractors in the parameter space
d as a function of a; $T^3(\triangle)$, $T^2(x)$, locking (o), and chaos (□). a) b = 0.1; b) b = 0.5.

linear (or coupled) region where T^3 is dominant. We also note that the ratio of chaos shows a significant change at $a \sim 0.85a_c$, where lockings to cycles also increase, which seem to be essential to the appearance of chaos.

(4) The ratio of lockings to T^2 increases at $a \sim 0.7a_c$, but it decreases for $a > 0.9a_c$, where lockings to cycles increase rather rapidly. The ratio of T^2, however, does not vanish for $a \sim 1.1a_c$.

(5) If the coupling term (b) is large, lockings into T^2 are more common for the same nonlinearity. The increase of chaotic attractors also starts at a less nonlinearity for the system with a stronger coupling.

In Figs. 6.10, some examples of attractors are shown. We note the oscillatory behavior of T^2 and localized attractors of chaos. Lockings exist at the parameter values close to the values for the chaos. The chaos has a large measure near the points (x_i, y_i) $(i = 1,2,\cdots,p)$ which are the periodic points of the lockings (with the period p) at the nearby parameter.

It is important to note that the attractor of the map (6.4.1) is not always unique even for $a < a_c$ (for the circle map (2.2.1), it is unique for $A < A_c$). We observed the coexistence of two or three types of cycles and the coexistence of chaos and a cycle, for example at $a = 0.95a_c$. Thus, the resonance overlapping has already occurred for $a < a_c$ (i.e., in the invertible region).

Recently, Grebogi, Ott, and Yorke[9] have investigated the coupled circle map with various couplings. They studied the map

$$x_{n+1} = x_n + C_1 + (2\pi)^{-1} P_1(x_n, y_n) \quad (\text{mod } 1)$$
$$y_{n+1} = y_n + C_2 + (2\pi)^{-1} P_2(x_n, y_n) \quad (\text{mod } 1)$$

Making use of the periodicity of the $P_{1,2}$ in (x, y), they expressed them as Fourier series,

$$P_\sigma(x, y) = \sum_{r,s} A^\sigma_{r,s} \sin (2\pi(xr + ys + B^\sigma_{r,s}))$$

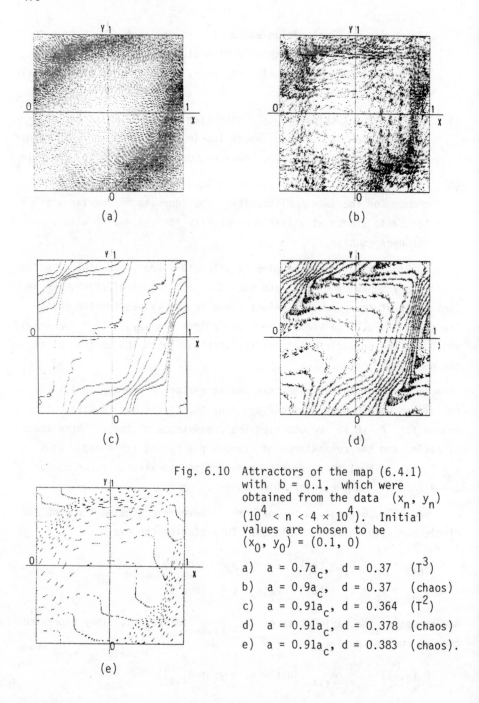

Fig. 6.10 Attractors of the map (6.4.1) with $b = 0.1$, which were obtained from the data (x_n, y_n) $(10^4 < n < 4 \times 10^4)$. Initial values are chosen to be $(x_0, y_0) = (0.1, 0)$

a) $a = 0.7a_c$, $d = 0.37$ (T^3)
b) $a = 0.9a_c$, $d = 0.37$ (chaos)
c) $a = 0.91a_c$, $d = 0.364$ (T^2)
d) $a = 0.91a_c$, $d = 0.378$ (chaos)
e) $a = 0.91a_c$, $d = 0.383$ (chaos).

where $j = 1,2$. They chose a set of A and B retaining only the Fourier mode with $(r, s) = (1, 0), (0, 1), (1, 1),$ and $(1, -1)$. Taking 512 random choices of (C_1, C_2) they calculated the ratio of T^3, T^2, locking, and chaos, for a given $\varepsilon/\varepsilon_c$, where ε_c is the value of ε at which the Jacobian of the map begins to have zero. Their criterion for the distinguishment of the attractors are again by Lyapunov exponents. The ratios for each attractor are shown in Table 6.II.

Table 6.II

Frequency of observation of different types of attractors for 512 random choices of (C_1, C_2) for $\varepsilon/\varepsilon_c = 3/16$ and 256 choices for each of the other values of $\varepsilon/\varepsilon_c$. In calculating the $\bar{h}$ we used 2×10^5 iterates for each random choice in the $\varepsilon/\varepsilon_c = 3/16$ case and 10^5 iterates for the other three values of $\varepsilon/\varepsilon_c$.

Type of attractor	$\varepsilon/\varepsilon_c = 3/16$	$\varepsilon/\varepsilon_c = 3/8$	$\varepsilon/\varepsilon_c = 3/4$	$\varepsilon/\varepsilon_c = 9/8$
Three-frequency quasiperiodic	92%	82%	44%	0%
Two-frequency quasiperiodic	8%	16%	38%	33%
Periodic	2%	2%	11%	31%
Chaotic	0%	0%	7%	36%

They also performed some numerical experiments for a 4-torus (three coupled circle maps). Their main results are as follows for $N = 3$ or $N = 4$:

   i) For small nonlinearity, an N-torus is rather common.

  ii) As the nonlinearity is increased it becomes less common.

 iii) Chaotic attractors are very rare for $N = 3$ for small to moderate nonlinearity, but are somewhat more common for $N = 4$.

  iv) There seem to exist such attractors that are chaotic but not strange, in the sense that they are apparently ergodic on the entire torus.

§6.5 Summary and Discussions

In the present chapter, we have investigated the fates of three-tori. Our results are summarized as follows:

i) Three-tori stably exist for a finite coupling. Various forms of the couplings are checked, but the conclusion does not change.

ii) As the coupling is increased, three-tori become feasible to phase-lock and two-tori appear. For the coupled delayed logistic maps, the lockings to two-tori become complete for a large coupling and a three-torus disappears.

iii) Lockings to two-tori from three-tori occur via tangent bifurcations. The ratio of two rotation numbers form a "double-devil's staircase" as a function of a bifurcation parameter.

iv) Phase instability in the high-dimensional torus can be investigated with the use of a coupled circle mapping. Results of the two-coupled circle mapping show the following features. The ratio of a three-torus in a parameter space is very low for weak nonlinearity. It increases as the nonlinearity is increased, but the most significant change is the increase of the measure of lockings.

Every simulation shows that the theory by Ruelle and Takens is not satisfactory. We know many systems which are structurally unstable, but are physically observed and important. Three-torus seems to belong to this category.

The experiment by Gollub and Benson[2] can be qualitatively explained from our results (see Fig. 6.11). Recently A. Libchaber et al.[4] have found an interesting phenomenon. They applied a horizontal magnetic field on the convection of mercury. Roughly speaking, the magnitude of a magnetic field corresponds to the strength of dissipation. When the magnetic field is strong (i.e., strongly dissipative system), the onset of chaos is shifted to the region of higher Rayleigh number, i.e., of larger nonlinearities.

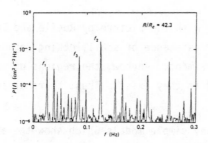

Fig. 6.11 Spectrum showing presence of three incommensurate frequencies, R/R_c = 42.3. All peaks are linear combinations of the three frequencies f_1, f_2 and f_3. (cited from Ref. 2); Prandtl number = 5.0, aspect ratio = 3.5).

For a small magnetic field, they observed the following scenario

cycle → torus → three-torus → locking → chaos,

which agrees with the experiment by Gollub and Benson[2] and with our simulation. Increasing the magnetic field ($B_0 \simeq$ 1500 Gauss), however, they found a new scenario, i.e., (see Fig. 6.12)

cycle → torus → chaos (= three ioncommensurate frequencies with exponential noise),

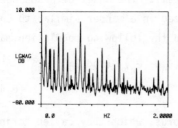 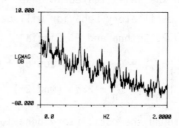

Fig. 6.12 Fourier spectra corresponding to the Ruelle-Takens-like scenario. R/R_c = 8.91 quasiperiodic state with frequencies f_1 and f_2 (vertical line indicate the position of future oscillator f_1); R/R_c = 8.98. Oscillator f_3 is present. Exponential noise is present (in a log-linear plot a constant slope in the recording). (cited from Ref. 4); Prandtl number = 0.025, aspect ratio = 4; magnetic field = 1500G)

which seems to agree with the picture by Ruelle and Takens[1]. Of course a possibility of the existence of small locking regimes cannot be excluded in the experiment. Anyhow, the regions of lockings are too small to be observed if they exist. Thus, there arises the following questions:

i) Are there any simple models which show the above behavior?

ii) Assuming that the models are found, what is the origin of the exponential noise?

iii) Are there any critical phenomena at this transition?

iv) What is the role of the strong nonlinearity for the above experiment for the observation of the new transition?

There has been some experiments which show the existence of 3- or 4- or 5-tori[5]. The behavior like Libchaber's experiment, however, has not been observed.

Some numerical works have also appeared quite recently, such as a 3- or 4-torus in a system of coupled differential equations[11] and a 3-torus in an optical bistable system[12].

In §§3 and 4, we have again encountered the oscillation of torus, which was investigated in Chap. 4 in detail. The onset of chaos due to this oscillatory behavior will be analyzed in a manner similar to Chap. 4. J.P. Sethna and E. Siggia[10] studied the following modulation map

$$\begin{cases} \theta_{n+1} = \theta_n + \omega + a \cos \theta_n + b \cos (\theta_n - \varphi_n) \quad (\text{mod } 2\pi) \\ \varphi_{n+1} = \varphi_n + 2\pi c \quad (\text{mod } 2\pi) \end{cases} \tag{6.4.1}$$

and found the oscillatory behavior of torus, which they called "crinkling". Oscillation of tori may be a generic scenario for the onset of chaos in a 3-torus system.

At the rest of the present section, some future problems on T^3 will be discussed.

(1) In the circle map for T^2, all tori are stable for $A < A_c$, which collapse simultaneously at $A = A_c$ and chaos appears for $A > A_c$.

Thus, a dissipative version of a KAM theory can be constructed. How about for T^3? There does not seem to exist some critical parameter a at which chaos appears simultaneously for various values of d. The construction of a dissipative version of KAM theory, therefore, is not trivial and will be an important problem.

We also note that the resonance overlapping gradually occurs even for $a < a_c$ for the model of T^3, while it appears only for $A > A_c$ for the circle map for T^2. Detailed study on the relation between the appearance of chaos and the resonance overlapping is left to the future.

(2) In the case of one-parameter lockings for T^2 in Chap. 2, a continued fraction expansion has been a very powerful method for the study of a devil's staircase. In T^3, lockings form a double devil's staircase. Is it possible to extend the method to the lockings in T^3?

(3) As is seen in the present chapter, the onset of chaos seems to occur from a locking to a cycle. In T^2, a period-doubling cascade from a locking is an essential mechanism for the onset of chaos. In the map (6.4.1) for $a < a_c$, period-doublings of lockings have not yet been observed and they do not seem to be important for the onset of chaos from T^3. Then, what is the relevant mechanism of the onset of chaos?

It is also important to characterize a chaotic orbit. The notions of ordering and disordering for T^2 are not easily applicable to T^3, because the rotation in the phase space (x, y) prevents a simple extension of the notions to a T^3-system, and some new ideas will be necessary. Also, the numerical study on the power spectra for the chaos in the coupled circle map (6.4.1) has to be performed.

(4) In connection with the experiments, it will be necessary to construct a "physical" theory for T^3 and chaos. Chaos might exist in the region with small nonlinearity a. It takes, however, a very long time to distinguish chaos from T^3 and the effect of noise in real systems prevents us from detecting chaos. Thus, we have to

construct a theory for the observability of chaos near T^3, which includes both the time for the observation and the effect of noise.

(5) T^3 stably exists. How about T^4, T^5, etc.? There is no reason to doubt their existence. The lockings to lower-dimensional tori and to cycles may perhaps play an essential role in such systems, which justify the recent success of a theory of a low-dimensional dynamical system for the onset of chaos. To study such a high-dimensional torus, the author has recently studied the following coupled circle map

$$x_{n+1}(i) = x_n(i) + a \sin (2\pi x_n(i)) + d + b (\sin (2\pi x_n(i + 1))$$
$$+ \sin (2\pi x_n(i - 1)) - 2 \sin (2\pi x_n(i))) \ ,$$

$$(6.4.2)$$

where $i = 1,2,\cdots, N$ denotes a spatial coordinate of a one-dimensional lattice. The map shows a variety of propagating spatial patterns, some of which will be discussed in Chap. 7.

There remains a lot of problems on the chaos from T^3, which will hopefully be solved in the future.

REFERENCES

*) The contents of this chapter are based on K. Kaneko, Prog. Theor. Phys. 71 (1984) 282; see also K. Kaneko, in The Theory of Dynamical Systems and Its Applications to Nonlinear Problems, (World Sci. Pub., 1984, ed. H. Kawakami).

1. Ruelle and Takens, Comm. Math. Phys. 20 (1971) 167. See also S. Newhouse, D. Ruelle, and F. Takens, Comm. Math. Phys. 64 (1978) 35.

2. J.P. Gollub and S.V. Benson, J. Fluid Mech. 100 (1980) 449.

3. J. Maurer and A. Libchaber, J. Phys. Lett. 41 (1980) L515.

4. A. Libchaber, S. Fauve and C. Laroche, Physica 7D (1983) 73.

5. M. Gorman, L.A. Reith and H.L. Swinney, Ann. N. Y. Acad. Sci. 357 (1980) 10; R.W. Walden, P. Kolodner, A. Passner and C.M. Surko, Phys. Rev. Lett. 53 (1984) 242; K.R. Sreenivasan, in Fundamentals of Fluid Mechanics, (Springer, to be published).

6. H. Yahata, Prog. Theor. Phys. 69 (1983) 396.

7. O.E. Rössler, Phys. Lett. 71A (1979) 155.

8. I. Shimada and H. Nagashima, Prog. Theor. Phys. 61 (1979) 1605.

9. C. Grebogi, E. Ott, and J.A. Yorke, Phys. Rev. Lett. 51 (1983) 339, and Physica 15D (1985) 354.

10. J.P. Sethna and E.D. Siggia, Physica 11D (1984) 193.

11. R.K. Tavakol and A.S. Tworkowski, Phys. Lett. 100A (1984) 65 (3-torus) and 273 (4-torus).

12. P. Davis and K. Ikeda, Phys. Lett. 100A (1984) 455.

Chapter 7

TURBULENCE IN COUPLED MAP LATTICES

Fishes and Scales
by *M. C. Escher*

§7.1 Introduction

Recent studies on low-dimensional dynamical systems have made great advances, which elucidate various aspects of chaos and the mechanism of its onset, as we have seen in the previous chapters. The success of the low-dimensional theory is, however, limited to systems with a few number of excited modes, which are relevant near the onset of turbulence owing to Ruelle-Takens' picture[1] and abundance of phase lockings. For the fully developed turbulence, however, the number of excited modes must be very large. Then the following questions arise; what happens in a system with a large number of excited modes and with a spatial complexity? Can the low-dimensional chaos be an elementary process for the turbulence? How are the spatial patterns in complex systems characterized? What is the effect of spatial patterns on the onset of chaos? Such questions are important for the study of fluid turbulence[2], chemical turbulence[3], optical turbulence[4], nonlinear field theory[5],[6], pattern formation problems, neuron network systems[7], and so on. The main topics are as follows; characterization of patterns, bifurcations of a solution with a spatial structure and transitions to chaos, characterization of spatial complexity, estimation of the fractal dimensions and Lyapunov spectra[8],[9], propagation of patterns and information, spatial bifurcation, stability of a direct product state, and phase transitions in spatial structures, and so on.

Another important topic in nonlinear system is "soliton", in the general sense[10], that is a topological excitation in nonintegrable systems. Some well-known examples are a kink, a vortex, and a breather. Here we extend the notion of a kink to dissipative systems and study the effect of chaos on the kink-antikink patterns.

In the present chapter we report some results on such questions by making use of coupled map lattices[11] introduced by the author. The coupled map lattice is given by the dynamical system with a discrete time (mapping) on lattice points. The models we study in the present chapter are given by

$$(I) \quad x_{n+1}(i) = f[x_n(i)] + D\{[x_n(i + 1) + x_n(i - 1)]/2 - x_n(i)\}$$

(II) $\quad x_{n+1}(i) = f[x_n(i)] + D'\{[f(x_n(i+1)) + f(x_n(i-1))]/2 - f[x_n(i)]\}$

where $i = 1, 2, \cdots, N$ denote one-dimensional lattice sites with periodic boundary condition $(x_n(N+1) = x_n(1))$. (In Secs. 4 and 5 we consider some other versions.) The function $f(x)$ is chosen to be $1 - Ax^2$ (i.e., coupled logistic lattice) or $x + A \sin(2\pi x) + C$ (coupled circle lattice).

We note that the model (I) corresponds to the "past coupling" in the sense that the coupling is given by the values $x_n(i)$'s at the previous step n, while the model (II) corresponds to the "future coupling" in the sense that the coupling is given by the values $f(x_n(i))$'s which are the values of $x_{n+1}(i)$'s (i.e., at the future step) in the absence of the coupling term. The future coupling model has an advantage, since it is close to a continuum model (thus, it can be a model for the partial differential equation system) and the zigzag instability is harder to occur (see §7.3).

§7.2 Period-doublings of Kink-antikink Patterns

One characteristic pattern of the coupled logistic lattice (CLL) ((I) and (II)) appears for the parameter region A where the period-doubling bifurcations proceed for the logistic map. The pattern is characterized by flat regions and domains (kinks or antikinks). The phase of the periodic oscillation (with a period 2^n) is the same in a flat region and it differs by flat regions. An example of the pattern is shown in Figs. 7.1-7.4, where the initial condition is given by $x_0(i) = \sin(2\pi i/N)$ and the period is 2 (Fig. 7.1), 4 (Fig. 7.2), 16 (Fig. 7.3) respectively while the attractor is chaos in Fig. 7.4. As A is increased, period-doubling bifurcations proceed, which lead to chaos. The process of period-doublings is essentially the same between the models (I) and (II). We note the following features:

(a) Doubling occurs only a finite number of times in general. The number can depend on initial conditions, coupling D, and the size N.

(b) The width of a kink decreases as A is increased from the bifurcation point. For example the width is 42 ($A = 0.752$), 27 ($A =$

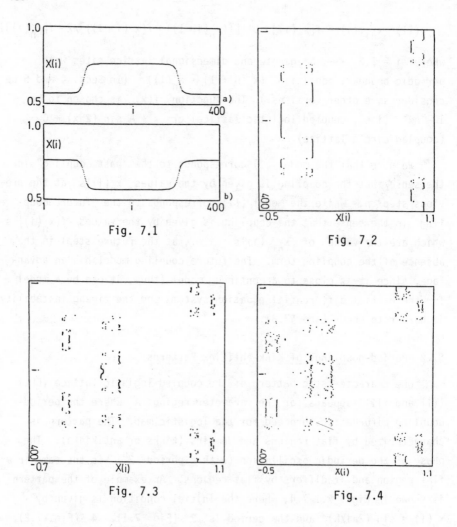

Fig. 7.1

Fig. 7.2

Fig. 7.3

Fig. 7.4

Figs. 7.1-7.4   Snapshot of the CLL (II) with the initial condition
$x_0(i) = \sin(\pi i/N)$ with $N = 400$ and $D = 0.05$,
after the transients are discarded. The value A
is chosen to be 0.755 (Fig. 7.1), 1.3 (Fig. 7.2),
1.395 (Fig. 7.3), and 1.43 (Fig. 7.4). The
period of the attractor is 2 for Fig. 7.1 (the
pattern repeats a and b alternatively), 4
(Fig. 7.2), 16 (Fig. 7.3), while the attractor is
chaos for Fig. 7.4.

0.755), 18 (A = 0.76), 8 (A = 0.8), and 4 (A = 0.9) for the model (II) with N = 200, D = 0.05 and with the initial condition $x_0(i) = \sin(\pi i/N)$. (Here we regard that a site belongs to a kink if the value $x(i)$ differs more than 0.02 from a flat region; we note that the kink structure with period two appears at $A \gtrsim 0.75$.) If we consider the continuum limit, the form of a kink obeys the equation

$$-Dd^2/dr^2 x(r) = F(x) = (Ax^2 + x - 1)(A^2 x^2 - Ax - A - 1)$$
$$= A^3(x - x_0)(x - x_0')(x - x_+)(x - x_-)$$

where x_0 and x_0' are unstable fixed points for the logistic map while x_+ and x_- are two stable points with period two for the map. The width of a kink L, then, obeys

$$L \propto (A - 0.75)^{-1/2} D^{1/2}$$

at $A \gtrsim 0.75$. The relation agrees with the numerical results.

(c) Period-doubling occurs not only in "time" but also in "space", if the kink-antikink patterns exist. Period-doubling brings about the spatial structure with more complexity. Structures with smaller wavelengths appear owing to the doubling. Small spatial structures are feasible to disappear as the increase of the coupling (i.e., a structure with a fewer kinks appears). See, e.g. Fig. 7.4, where small kink-antikink structures cannot be seen.

(d) Even after the transition to chaos occurs, patterns of kink-antikinks are preserved and chaos is localized in each domain. As the nonlinearity is increased further, the structure collapses, by the "domain merging". See Fig. 7.4 for the 4-band structure. The localization of chaos is also confirmed from the Lyapunov vectors, which will be discussed in §7.7.

(e) Many attractors coexist for the CLL with large N. They are obtained by the change of the initial condition $x_0(i)$. If the coupling vanishes, k^N attractors coexist for the parameter region where a stable cycle with period k exists for the one-dimensional mapping. It seems that the number of the attractors in the presence of the coupling is at least of the same order. Some of the attractors are characterized by the number and the position of kinks. The number of the attractors decrease as the coupling is increased since the width of a kink becomes larger.

The increase of the number of the attractors by the period-doubling may imply that our dynamical system can be an information source.

§7.3 Zigzag Instability and Transition from Torus to Chaos

Another interesting pattern in CLL is an "antiferro-like" structure, which is shown in Fig. 7.5 (the figure shows a cycle with period two). It is characterized by an alternate structure with a wavelength two, which appears through the "zigzag" instability. This structure is remarkably seen in the model (I) though it can be seen in a small parameter region for the model (II). We note that there can be kinks in the antiferro-like structure as can be seen in Fig. 7.5, which is rather analogous to solitons in polyacetylene[12]. One kink is clearly shown in

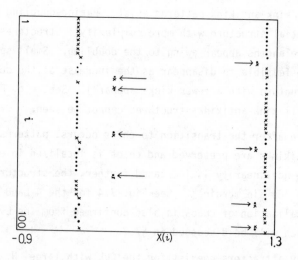

Fig. 7.5 Snapshot of Map (I) with the initial condition x(i) = sin (πi/N) and A = 0.74 and D = 0.2. The symbols x and ● denote the odd and even sites respectively. We note the existence of some kinks (shown by arrows). The attractor is a 2-cycle.

Fig. 7.6, where the parameter is close to the onset of zigzag instability and the width of a kink is large. As the nonlinearity A is
increased, there occurs a Hopf bifurcation and a torus appears (see
Fig. 7.7). For the attractor in x(1) - x(2) space, see Fig. 7.8a).

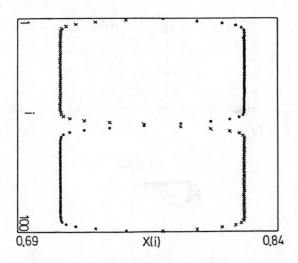

Fig. 7.6 Kink-antikink pattern for Map (I) with A = 0.392
         and D = 0.2. The symbols x and ● denote the
         odd and even sites respectively.

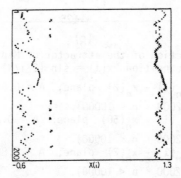

Fig. 7.7 Snapshot of the CLL (I) with the initial condition
         $x_0(i) = \sin(2\pi i/N)$ with N = 200, after the
         transients are discarded, D = 0.2, and A = 0.8.
         Even sites are denoted by x, while odd ones are
         shown by ● .

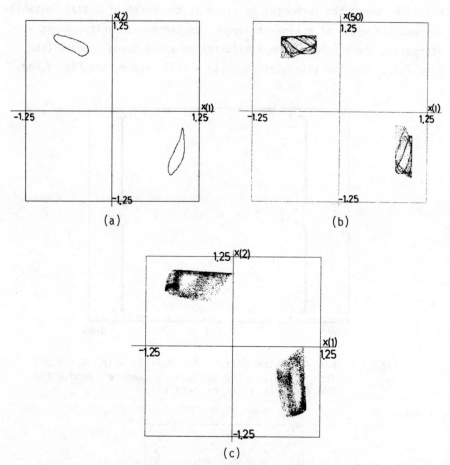

Fig. 7.8 Projection of the attractor of Map I, with the initial condition x(i) = sin (2πi/N) and D = 0.2.

a) $x_n(1) - x_n(2)$ plane; A = 0.84
    (2000 < n < 10000).
b) $x_n(1) - x_n(50)$ plane; A = 0.84
    (2000 < n < 10000).
c) $x_n(1) - x_n(2)$ plane; A = 0.88
    (2000 < n < 10000).

The shape of the projected torus (Fig. 7.8a for example) differs by the projected space (see for example Fig. 7.8b) for the projection into $x(1) - x(50)$ space). As the nonlinearity is increased further, the torus, is modulated (3-torus is expected) and finally chaos appears (see Fig. 7.8c). We note the following features:

i) We can construct various types of structures in which the number of kinks differ, by choosing suitable initial conditions. Kinks cause the modulation of the phase of the torus, which bring about the transition to chaos.

ii) In some parameter regions, two types of patterns can coexist, i.e., the structures in §2 and the present section. Small irregularities in the pattern in §2 can cause the transition to the antiferro-like pattern (e.g., the pattern in Sec. 2 with some singularity or the initial condition with some singularities (e.g., $x(i) = \sin(\pi i/N)$) show this type of behavior).

iii) Even after the transition from torus to chaos, the kinks can exist, which do not change their positions by the iteration of the map. As the nonlinearity is increased further, the antiferro-like structure collapses and the chaotic structure with more spatial complexity appears.

iv) The antiferro-like structure is also observed in other parameter regions of A, where basic dynamics shows not a 2-cycle but a 4-cycle or chaos. (See for an example Fig. 7.4.)

§7.4 Spatiotemporal Intermittency

Another important aspect for the coupled map lattice is spatio-temporal intermittency.

Intermittency has been used in two different meanings; i.e., one for the intermittency by Pomeau and Manneville[13], where the notion is purely temporal in low-dimensional dynamical systems, and the other for the original meaning in fluid mechanics[14] where the spatiotemporal self-similar structure of eddies are important. It has not yet been clear whether the above two notions are related or not. Though our spatiotemporal intermittency cannot establish the relationship, it may give an insight towards the understanding of it, in the sense that our model starts from Pomeau and Manneville's intermittency and shows a geometrical structure in spacetime.

(I) Models

As models for the spatiotemporal intermittency, we consider the coupled map lattice

$$x_{n+1}(i) = f(x_n(i)) + \varepsilon\{g(x_n(i + 1)) + g(x_n(i - 1)) - 2g(x_n(i))\} ,$$

$$(7.4.1)$$

where the one-dimensional map $x_{n+1} = f(x_n)$ shows a stable periodic cycle but the stability is weak and Pomeau and Manneville's inter-mittency[13] occurs at the nearby parameters. The model we studied numerically are as follows:

(A) Coupled Circle Lattice

$f(x) = x + A \sin 2\pi x + C$ (mod 1)

$g(x) = \sin 2\pi x$

where $A = 0.2$ and $C = 0.55$, at which a stable cycle with period two $(x_2^* = f(x_1^*),\ x_1^* = f(x_2^*)$ exists for the one-dimensional map (see Fig. 7.9a)).

(B) Coupled Logistic Lattice

$f(x) = 1 - Ax^2$

$g(x) = f(x)$

where A is chosen to be 1.752, which is slightly larger than the onset value for the appearance of a stable three-cycle via the inter-mittent transition $x_2^* = f(x_1^*)$, $x_3^* = f(x_2^*)$, and $x_1^* = f(x_3^*)$ (see Fig. 7.9b)).

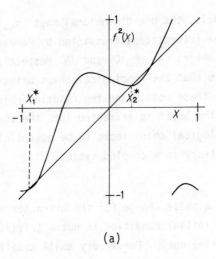

(a)

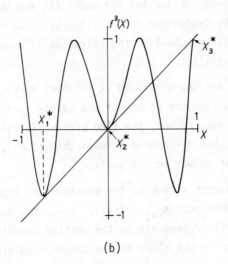

(b)

Fig. 7.9a) Twice-iterated circle map $x_{n+1} = f^2(x_n)$ $((A))$
with $A = 0.2$ and $C = 0.55$.

b) Three-times iterated logistic map $x_{n+1} = f^3(x_n)$
$((B))$ with $A = 1.752$.

In these models, the one-dimensional maps $x_{N+1} = f(x_n)$ are chosen so that the intermittent transition by Pomeau and Manneville[13] occur if the parameters (A or C and A respectively) are changed slightly. We note that the topological chaos exists in the dynamics $x_{n+1} = f(x_n)$ for these models and the chaotic orbit appears as a transient before the orbit is attracted into the stable cycle. The existence of topological chaos seems to be essential for the spatio-temporal intermittency in a coupled system.

(II) Phases

There occurs a phase change for the attractor as the coupling is increased, if the initial condition is not a trivial one (e.g., $x(i) =$ constant is a trivial one). For a very small coupling the pattern is essentially the same as in the case without coupling. Thus, the attractor is a cycle (with period two for model (A) and three for (B)), though the phases of the cycles can differ sites by sites for the last two models. For a larger coupling the attractor is chaotic, where bursts exist spatiotemporally.

The bursts are characterized as follows; first, the value $\dot{x}(i)$ deviates largely from the values of the stable periodic points for the one-dimensional map; second, the time series of $x(i)$ shows a chaotic behavior; third, the difference between the values $x(i)$'s for the neighboring sites do not remain small.

The third aspect is due to the existence of topological chaos for the one-dimensional map $x_{n+1} = f(x_n)$, i.e., the map $x_{n+1} = f(x_n)$ shows the sensitive dependence on the initial condition in some regions of x, where x is not close to the periodic points x_i^*.

To make the third point clear, we consider the initial condition $x(i) = x_1^*(1 < i \leq N/2)$ and $x(i) = x_2^*(N/2 < i \leq N)$ for the models (B) and (C). When the nonlinearity in one-dimensional map $x_{n+1} = f(x_n)$ is small and there is no topological chaos, kink-antikink patterns exist stably, which can typically be seen in Fig. 7.1.

In our case, this pattern cannot exist stably, because in the region of a kink or antikink, $x_n(i)$ can take the value at which the one-dimensional map $x_{n+1} = f(x_n)$ shows a transient chaotic behavior. Thus, the kinks or antikinks become bursts which propagate if the coupling is large enough. As an example, see Fig. 7.10, for the snapshot of the model (A).

The "phase transition" from a direct product state to a burst state is not simple at all in the parameter space. We note the following aspects:

(i) The transition occurs not at one point in the parameter space but at many points. That is, for $\varepsilon > \varepsilon_0$ we have a burst state, and for $\varepsilon_0 > \varepsilon > \varepsilon_1$ we have a laminar state, and for $\varepsilon_1 > \varepsilon > \varepsilon_2$ again a burst state, and so on. As we study in more detail, the parameter space, there appears more transition points. The structure here was found in the parameter space for a fixed initial condition, though this type of a fine structure was also observed in basin structures[15)-17)], and in the velocity space of colliding kinks in a ϕ^4 system[18)].

(ii) Near the transition region, we have observed some complicated phases for the models (A) and (B). Examples are as follows:

a) kinks and antikinks are recombined and a homogeneous phase appears.

b) localized chaos; the burst regions exist but they cannot propagate into the whole space and are confined within some regions. Examples are shown in the next section.

(III) Visualization — Cellular Automaton Reductions

Here we give some pictures for the patterns of bursts and laminar clusters. Patterns in the model (A) are shown in Figs. 7.11 where initial conditions are chosen to be $x(i) = x_1^*$ (for $1 < i \leq N/2$) and $x(i) = x_2^*(N/2 < i \leq N)$. That is, bursts exist initially only at sites $i = N$ and $i = N/2$. Here, the following reduction method to two-state cellular automata is used; $S(i) = 1$ if $|x(i + 1) - x(i)|$ (mod 1) $> \delta$

198

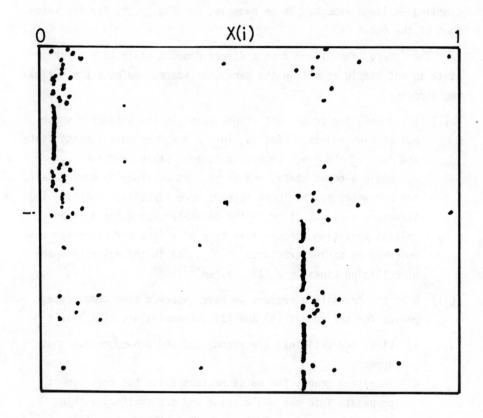

0　　　　　　　　　　　　　X(i)　　　　　　　　　　　　　1

Fig. 7.10　Snapshot for $x_n(i)$ for model (b) with $\varepsilon = 0.024$
from the initial condition $x_0(i) = \sin(2\pi i/N)$,
after the transients have decayed out.

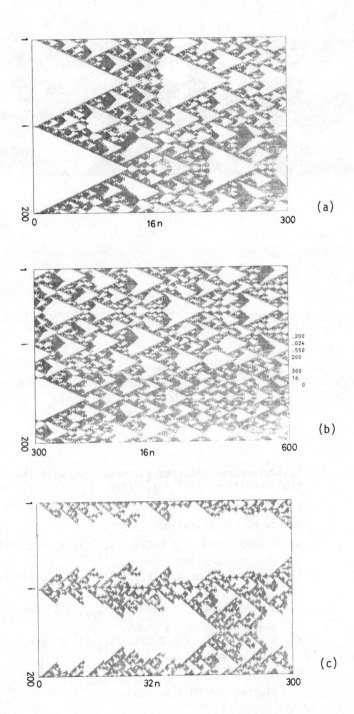

.200
.024
.550
200

300
16
0

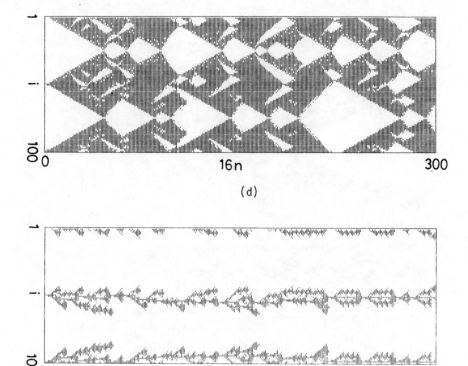

(d)

(e)

Fig. 7.11 Spatiotemporal patterns for model (A) with the initial condition $x(i) = x_1^*$ (for $1 < i \leq N/2$) and $x(i) = x_2^*$ (for $N/2 < i \leq N$). See text for the method of visualization.

a) $N = 200$ and $\varepsilon = 0.024$. $x_{16n}(i)$'s are plotted for $0 < n < 300$.

b) Continued from a). $x_{16n}(i)$'s are plotted for $300 < n < 600$.

c) $N = 200$ and $\varepsilon = 0.0238$. $x_{32n}(i)$'s are plotted for $0 < n < 300$.

d) $N = 100$ and $\varepsilon = 0.025$. $x_{16n}(i)$'s are plotted for $0 < n < 300$.

e) $N = 100$ and $\varepsilon = 0.0236$. $x_{32n}(i)$'s are plotted for $0 < n < 300$.

(δ = 0.05) and S(i) = 0 otherwise. For the figures, the sites with S(i) = 1 are shown by dots. Figure 7.12 gives the pattern for the same model with the same reduction, but with the random initial condition (x(i) = random number homogeneously distributed in (0, 1)).

Some patterns for the model (B) are shown in Figs. 7.13, where initial conditions are again random or $x(i) = x_1^*$ (for $1 < i \leq N/2$) and $x(i) = x_2^*$ (for $N/2 < i \leq N$) and the criterion for the burst is given by $|x(i + 1) - x(i)| > 0.1$.

We note the following aspects:

(1) The geometrical pattern of burst and laminar states: Clusters of laminar sites with various sizes form a structure with a similarity, which is quite analogous to the patterns in cellular automata found by Stephen Wolfram[19] (see also §7.6). The difference between coupled map lattices and cellular automata lies in the complexity of an element itself in the former models. So, it is rather astonishing that the reduction to two-state cellular automata works rather well for the present case.

(2) Localized chaos: As has been described in (II), the burst remains in a finite space for some parameter regions. Here we note that the pattern looks quite similar, like the growth of dendritic crystals.

(3) Self-organization from random initial configurations: The simulations from the random initial configuration give similar figures to the corresponding patterns in Fig. 7.11. That is, the geometrical structure or dendritic-crystal-like pattern appears also from random initial configurations. Thus, these patterns are inherent to our dynamical system, not due to the special choice of the initial conditon.

(4) The patterns in the model (B) are not so beautiful as those for the model (A), in the sense that the geometrical pattern with self-similarity cannot be clearly seen. In the patterns in (B), the propagation is not so straight and it moves like Brownian

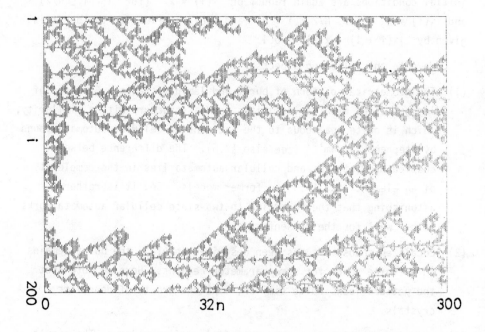

Fig. 7.12 Spatiotemporal patterns for model (A) with the random
initial condition, i.e., $x(i)$ = homogeneous random
number in (0, 1). $x_{32n}(i)$'s are plotted for
$0 < n < 300$ with the use of the same method as in
Fig. 7.11 for visualization.

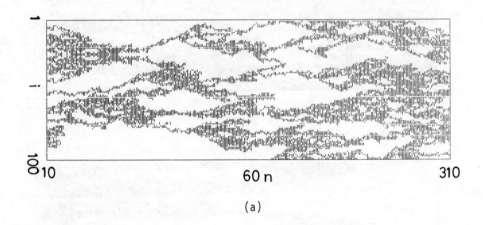

(a)

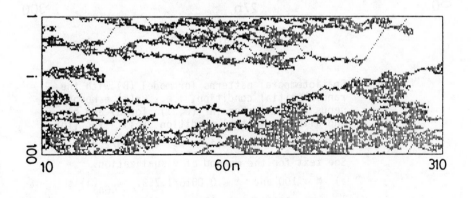

(b)

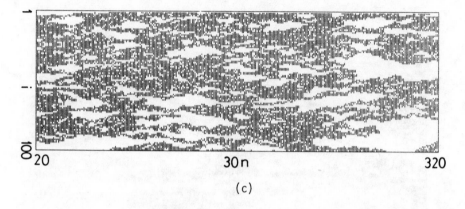

(c)

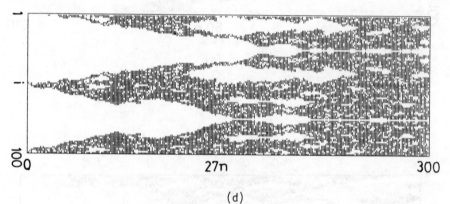

(d)

Fig. 7.13 Spatiotemporal patterns for model (B) with the random initial condition, i.e., $x(i) =$ homogeneous random number in $(0, 1)$ for a) and b) and with the initial condition $x(i) = x_1^*$ (for $1 < i \leq N/2$) and $x(i) = x_2^*$ (for $N/2 < i \leq N$). See text for the method of visualization.

a) $N = 100$ and $\varepsilon = 0.0018/1.752$. $x_{60n}(i)$'s are plotted for $10 < n < 310$.

b) $N = 100$ and $\varepsilon = 0.00105$. $x_{60n}(i)$'s are plotted for $10 < n < 310$.

c) $N = 100$ and $\varepsilon = 0.0011$. $x_{30n}(i)$'s are plotted for $20 < n < 320$.

d) $N = 100$ and $\varepsilon = 0.0021/1.752$. $x_{27n}(i)$'s are plotted for $0 < n < 300$.

motion, which is expected, since the original one-dimensional mapping has a topological chaos. In the model (A), the map $f^2(x)$ is monotone at most parts of x and $x_n(i)$ increases (without mod) monotonically as n for most of the time. Thus, the burst propagation in model (A) is more regular than in (B), which makes the pattern in (A) geometrically more regular.

(IV) Mechanism

A typical aspect of the patterns in (III) is that they are formed by the existence of two kinds of the propagation, i.e., that of laminar clusters and that of bursts. As an example, we consider the pattern like Fig. 7.14, which appears with various sizes in Figs. 7.11. The elementary pattern in Fig. 7.14 is formed by the propagation of laminar regions with speed v_L and that of burst regions with speed v_B. The ratio v_B/v_L is less than one in the patterns in Figs. 7.11, which goes to zero as ε is decreased. The propagation, of course, does not occur at a constant speed and the speed of the propagation has some fluctuations, though they are very small for model (A).

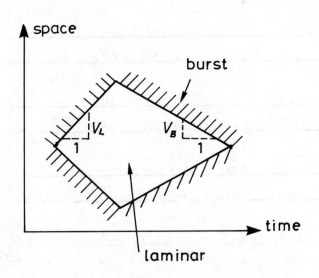

Fig. 7.14 Schematic representation for the elementary laminar cluster.

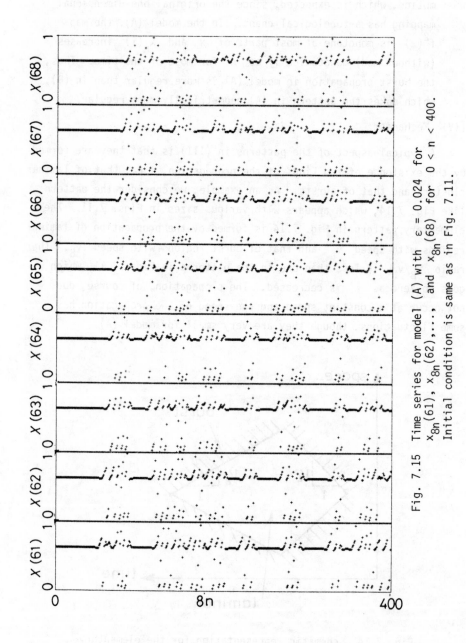

Fig. 7.15 Time series for model (A) with ε = 0.024 for
x₈ₙ(61), x₈ₙ(62),...., and x₈ₙ(68) for 0 < n < 400.
Initial condition is same as in Fig. 7.11.

The time series for the model (A) for a sequence of lattice points are shown in Fig. 7.15, which clearly represents the existence of the two kinds of propagation speeds.

What is the mechanism for the existence of two speeds? Here we discuss roughly their meanings. The map $f^p(x)$ ($p = 2$ for model (A) and 3 for model (B) has the form

$$f^p(x) - x_i^* \quad \alpha(x - x_i^*) + \beta(x - x_i^*)^2 \; ; \quad \alpha \sim 1$$

near some periodic point x_i^* (see Fig. 7.16). Assume that the sites $i \leq l_0$ belong to laminar regions I_L (close to x_i^*). The motion of $x(i_0)$ departs from the laminar region only if $x - x_i^*$ exceeds Δ by the effect of $g(x_n(i + 1))$. Thus, the propagation of a burst occurs if

$$\frac{\varepsilon}{2} \sum_{k=1}^{L} g(x_k(i_0 + 1)) - g(x^*) \sim \Delta$$

The speed v_B is related with the coupling term, while the speed v_L is related with the probability that a chaotic orbit falls on the region I_L.

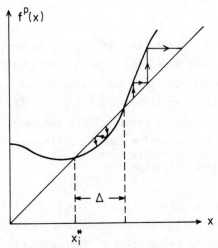

Fig. 7.16 Schematic representation for $f^p(x)$ near $x = x_i^*$.

As ε is decreased, the propagation speed of burst decreases till it vanishes at $\varepsilon \sim \varepsilon_c$. The decrease of v_B is roughly given by $(\varepsilon - \varepsilon_c)^\nu$ with $\nu = (0.6 \sim 0.9)$. See Fig. 7.17 for example, for the

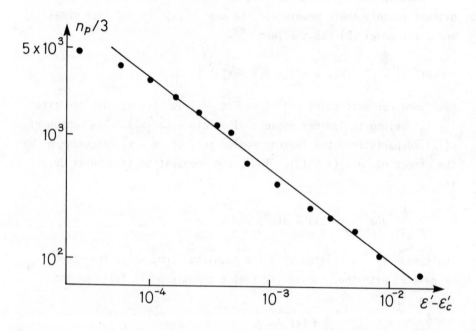

Fig. 7.17 The propagation time n_p as a function of coupling ε for model (B) with $N = 100$ with the initial condition $x(i) = x_1^*$ (for $1 < i \leq N/2$) and $x(i) = x_2^*$ (for $N/2 < i \leq N$). The notation $\varepsilon' = \varepsilon A$ is used. ε_c' is estimated to be 0.001845. The propagation time is defined as the time step when the bursts from $i = 0$ and $i = 50$ make the first collision.

behavior of the propagation time ($\propto N/v_B$) for the model (B), which gives the exponent $\nu \sim 0.75$. The exponent ν seems to depend on the models. The accurate calculation of the exponent, however, is rather hard, since the transition does not occur at one point and some data have large deviations from the above power law fit.

Another interesting quantity is the distribution of the size of laminar clusters. In the case of a class-3 cellular automaton[16], the distribution function of 0-sites decays exponentially if it starts from the disordered configuration while it shows a power-decay if it starts from a configuration with $x(i) = 1$ for $i = i_0$ and 0 otherwise. (See §7.6 for cellular automata).

Instead of the direct calculation of the distribution function of clusters, the "laminar sequence density" $Q(n)$ is numerically calculated for our model, which is defined as the spatial sequence of exactly n adjacent laminar sites, bordered by sites with bursts. The density $Q(n)$ and the distribution of the laminar cluster $T(n)$ is related by

$$Q(n) \propto \sum_{i=n}^{\infty} (2T(i)/i) \ .$$

According to the numerical results, the density $Q(n)$ (thus, $T(n)$ also) decays exponentially even if the initial configuration is $x(i) = x_1^*$, $(i \leq i_0)$ and $x_2^*(i > i_0)$ (see Fig. 7.18). The rate of the decay decreases as ε decreases and at some values of ε close to ε_c, some data show power-decay-like behavior, though, again, the detailed study of the critical phenomena is rather hard.

(V) Lyapunov Spectra

Numerical results for the calculation of the Lyapunov spectra are shown in the section. Here we chose the initial condition $x(i) = 0.2 \sin (2\pi i/N)$ and iterated the map (B) with $N = 50$ by 9000 times after dropping the initial 10000 times of transients. The method of calculation is based on ref. 20). See references 8) and 9) for the Lyapunov spectra in high dimensional systems. The histograms for the Lyapunov exponents are shown in Fig. 7.19, where the numbers of the exponents within $[-0.37, -0.32)$, $[-0.32, -0.27), \cdots, [0.38, 0.43)$ are plotted, for $\varepsilon' = 0.0019, 0.00197, 0.0021, 0.0023$, and 0.0025 where $\varepsilon' = \varepsilon A = 1.752\varepsilon$.

As ε is increased, the number of positive exponents increases. Here we note that the histogram has a peak at $\lambda = \beta \sim -0.32$ and $\lambda = \alpha \sim 0.40$. As ε is increased, the number at $\lambda \sim \beta$ decreases and the

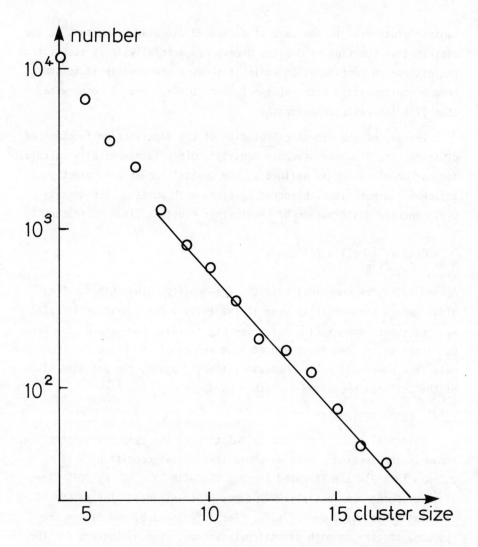

Fig. 7.18 Cluster distribution for the model (A) with
ε = 0.024 and N = 50 and the same initial

condition as in Fig. 7.11 $\sum\limits_{i=0}^{3} Q(4n + i)$ is

plotted as a function of laminar size n.
Q(n) is calculated from the data $x_{2m}(i)$
for $10^4 < m < 1.5 \times 10^4$.

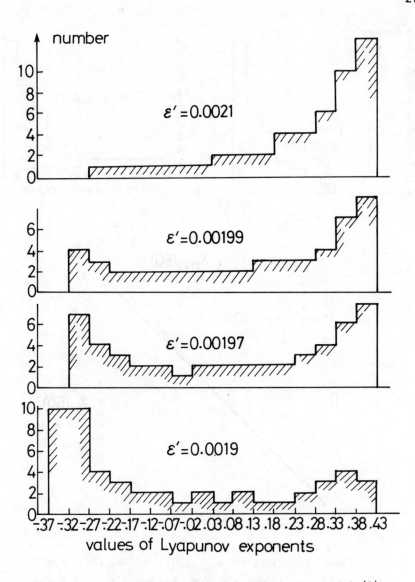

Fig. 7.19 Histogram of Lyapunov spectra for model (B) with N = 50. The number of Lyapunov exponents which take the values in the interval $(-0.37 + 0.05 \times j, -0.32 + 0.05 \times j)$ is plotted as a function of j. Initial condition is chosen as $x_0(i) = 0.2 \sin(2\pi i/N)$.

$\varepsilon' = \varepsilon A$. See text for the method of calculation.

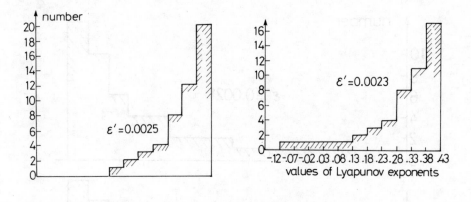

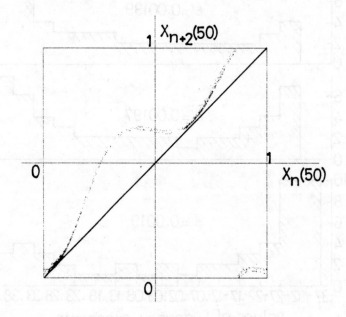

Fig. 7.20 Plots of $x_{2n+2}(50)$ vs. $x_{2n}(50)$ for
$1000 < n < 4000$ for model (A). N = 100
and $\varepsilon = 0.024$. Similar behavior is
obtained for model (B).

one at $\lambda \sim \alpha$ increases, while the number at $\lambda \sim 0$ remains small. The number $\lambda \sim \beta$ agrees with the exponent for the stable cycle with period three for the one-dimensional map $x_{n+1} = f(x_n)$. Thus, the Lyapunov exponent $\lambda \sim \beta$ corresponds to the laminar part of the motion. On the other hand, the exponent $\lambda \sim \alpha$ is related to the burst motion in the sense as follows: The data $(x_n(i), x_{n+1}(i))$ from the coupled map lattice equation lie close to a curve $x_{n+1} = F(x_n)$, where $F(x)$ is a function a little bit different from $f(x)$. Of course, the data scatter around $x_{n+1} = F(x_n)$ with some fluctuations. The function $F(x)$ is a sort of a "mean field mapping", in the sense that the average motion of the coupling term $\varepsilon/2\{g(x_n(i+1)) + g(x_n(i-1)) - 2g(x_n(i))\}$ gives a correction $F(x_n(i)) - f(x_n(i))$. The mapping $x_{n+1} = F(x_n)$ does not have a stable cycle and it gives a positive Lyapunov exponent about $\lambda \sim \alpha$ (see Fig. 7.20 for the numerical plot for $F^2(x)$). The fact that many exponents concentrate at $\lambda \sim \alpha$ for large couplings means that the turbulent motion is approximated by the direct product state of the one-dimensional map $x_{n+1} = F(x_n)$, though, of course, some fluctuations around the state exist.

The increase of the number of positive Lyapunov exponents is a process of the development of chaos into higher-dimensional one, though the dimension itself is not calculated here. The number of positive Lyapunov exponents shows a behavior $(\varepsilon' - \varepsilon'_c)^a$ with $a \sim (1/3 \sim 1/2)$ where ε'_c is the onset value for the chaotic motion (since the transition does not occur at one point, accurate calculation is impossible, but the fine structure is seen only in a small range of the parameter and a rough estimate for ε_c is possible).

The dependence of the number of positive exponents on the bifurcation parameter near the onset of chaos is also studied for the cases of the period-doubling and the transition from torus to chaos for the coupled logistic lattice, which shows the similar behavior $((\varepsilon - \varepsilon_c)^a;$ $a \sim 1/2)$ for the increase of the number, which will be discussed in §7.6.

§7.5 Period-doubling in Open Flow

Though the turbulence in closed systems such as Bénard convection or Taylor vortices has been extensively investigated from the viewpoint of dynamical systems, studies of open flow systems from that viewpoint are rare[2),21)]. One interesting feature for an open flow system lies in the change of a structure of a flow as it goes downstream[22)], such as the growth of a disturbance or the development of Karman eddies. In that sense, "space" can be considered as a kind of bifurcation parameter for an open flow system. In the present section a coupled map lattice (CML) model with asymmetric coupling is used to consider such flow systems. As a model we consider

$$x_{n+1}(i) = f(x_n(i)) + \varepsilon\{\alpha f(x_n(i+1)) + (1-\alpha)f(x_n(i-1)) - f(x_n(i))\}$$

$$(7.5.1)$$

where $i = 1, 2, \cdots, N$ is a lattice site and $f(x) = 1 - ax^2$. In the present section we present some results for the case with one-way coupling $(\alpha = 0)$ and with the boundary condition $x(1) = $ fixed at the unstable fixed point of a single logistic map $x^* = (\sqrt{1 + 4a} - 1)/(2a)$, though essential features doe not change if the coupling is asymmetric with the boundary condition $x(1) = $ fixed and $x(N) = x(N - 1)$.

(I) Spatial period-doubling

As the lattice site is increased, spatial period doubling is observed for various parameters. The pattern may be described as follows: At lattice sites $i(1) < i < i(2)$, $x_n(i)$ is a period-two cycle (values of $x(i)$'s at that cycle can differ by lattice sites), at $i(2) < i < i(3)$, $x_n(i)$ is a period-four cycle and so on. At some parameters (a, ε), the spatial period doubling stops at some order 2^k and the system settles down to a cycle with the period 2^k for $i > i(k - 1)$. For other parameters, the time series of $x_n(i)$ shows a chaotic behavior after a finite number of spatial period-doublings (usually larger nonlinearity a or a smaller coupling ε gives a chaotic behavior). Once chaos is attained at some lattice points i_c, no periodic behavior reappears for $i > i_c$. Some examples are shown in Figs. 7.21a)-e), where the initial conditions are chosen to be

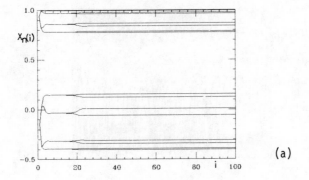

(a)

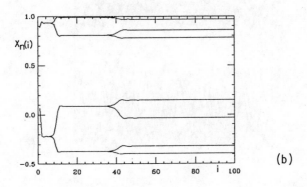

(b)

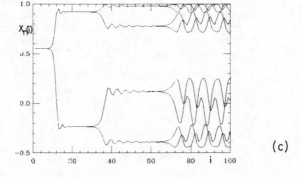

(c)

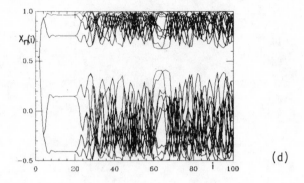

(d)

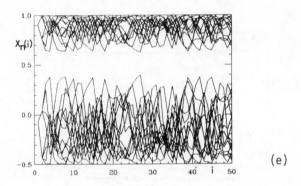

(e)

Fig. 7.21 Patterns by Eq. (7.5.1). x(i)'s are plotted
for n = 5001, 5002, •••, and 5020. Initial
conditions are x(i) = x* + 0.01, though the
pattern does not change so much if different
initial conditions are taken. The piecewise
lines denote $(i, x_n(i)) - (i + 1, x_n(i + 1))$.
System size N is 100 for a)-c) and 50 for d).

a)   a = 1.4   and   ε = 0.1
b)   a = 1.4   and   ε = 0.25
c)   a = 1.45 and   ε = 0.5
d)   a = 1.5   and   ε = 0.3
e)   a = 1.5   and   ε = 0.2
      (for n = 5001, 5002, •••, and 5025)

$x(i) = x^* +$ small disturbance, though the patterns do not depend on the initial conditions so much. The following points should be noted.

a)   The period-doubling is not caused by the change of a bifurcation parameter. The bifurcation occurs automatically as a lattice site goes downflow. In that sense, "space" plays a role of bifurcation parameter.

b)   No scaling relations are observed. As an example, let us consider a lattice point $i(k)$, at which the doubling from 2^{k-1} to 2^k occurs. The points $i(k)$ does not accumulate as k becomes large. On the contrary, the distance between $i(k + 1)$ and $i(k)$ becomes larger and larger as k becomes large. Scalings in a parameter space a or ε are not observed either, which is quite different from the commonsense in a low-dimensional chaos theory[23]. A possible reason for this is discussed in (IV).

(II) Mechanism

The mechanism of the above phenomenon seems to be rather simple. Assume that $x_n(i)$ is a cycle with period 2^k. Then the mapping at the site $i + 1$ is a logistic map modulated by the period 2^k. Then the amplitude of oscillation at the site $i + 1$ can become larger than that at the site i or a pitchfork bifurcation from 2^k to 2^{k+1} cycle occurs. In Figs. 7.22a)-c), $x_{n+1}(i)$ vs. $x_n(i)$ are plotted for $i = 5, 6, \cdots$, and 10. At these parameters in the figures, transition to chaos from 16-cycle occurs at the site $i = 6$, and the stochastic motion is propagated to the downstream.

(III) Lyapunov exponents and convective instability

In a system with a one-way coupling and the fixed boundary condition for the left end, the Jacobian for the CML (7.5.1) is a triangular matrix. Thus an eigenvalue of the product of Jacobians is given by the product of each diagonal element. Lyapunov exponents for the system (7.5.1), therefore, are easily calculated and each exponent has a one-to-one correspondence with the lattice site. Lyapunov exponent as a function of the lattice site is given in Fig. 7.23, where the parameters are same as for Figs. 7.22. We note that the exponent is positive only

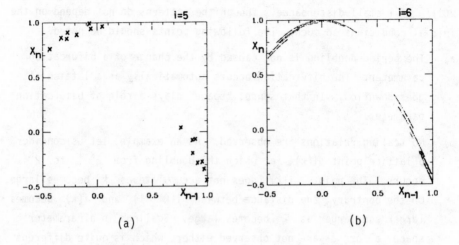

(a)

(b)

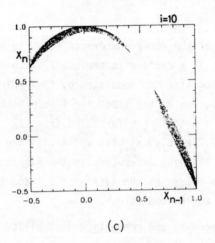

(c)

Fig. 7.22 $(x_{n-1}(i), x_n(i))$ is plotted for n = 5001,
5002, ... , and 7000 from Eq. (7.5.1) with
the same initial condition as for Fig. 7.21.
Lattice sites are i = 5 ((a); period-16),
i = 6 (b) and i = 10 (c).

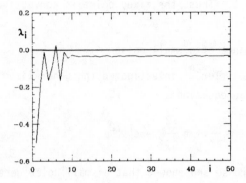

Fig. 7.23 Lyapunov exponent as a function of lattice site.
The parameters and the initial condition are same
as in Fig. 7.22. Calculations are performed by
10^5 iterations after 5000 times transients are
dropped.

at the site 6 and it approaches to some constant (negative) value for
the downstream. The stochastic motion at the downflow is essentially
due to the propagation of the turbulence of the upstream and cannot be
represented by the usual Lyapunov exponents. The convergence of the
exponent at the downstream means that the flow approaches some stationary state there.

In connection with the Lyapunov exponents, we have to be careful in
the difference between convective instability and absolute instability.
If the perturbation grows in a moving frame, it is called "convective
instability", while, absolute instability means the instability only in
a stationary frame[24]. In many cases our system shows a convective
instability. If the chaos arises only from the convective instability,
the notion of Lyapunov exponents in the co-moving frame is important[25].

As the simplest case, let us consider the stability of the homogeneous solution $x(i) = x^*$. From the calculation of Jacobian, it is
stable if $-1 < (1 - \varepsilon)f'(x^*) < 1$, i.e., $(1 - \varepsilon)(\sqrt{1 + 4a} - 1) < 1$.
The solution, however, is unstable for some co-moving frame if

$-1 < f'(x^*) < 1$. Thus, the fixed point is convectively unstable if $(\sqrt{1 + 4a} - 1) < 1$.

(IV) Importance of a small noise

Robert Deissler[24] investigated the generalized time dependent Ginzburg-Landau equation

$$\frac{\partial \psi}{\partial t} = a\psi - v \frac{\partial \psi}{\partial x} + b \frac{\partial^2 \psi}{\partial x^2} - c|\psi|^2\psi \qquad (7.5.2)$$

with a small noise and showed that a small noise generates a chaotic structure at the downstream. Thus, the microscopic noise plays an important role in the macroscopic dynamics for the system with convective (or spatial in his terminology) instability[9]. This is also true of our system. In a variety of cases, single and double precision calculations give different results, in the sense that the bifurcation from 2^k to 2^{k+1} occurs at different sites by precisions, if the 2^k-cycle is convectively unstable but not absolutely unstable ($i(k + 1)$ can be larger for the double precision). One reason that any scaling relations are not found lies in the above sensitive dependence on a small error in our system.

It is of importance to study the system (7.5.1) with a small noise added on every site. Numerical simulations of such systems show that (see Fig. 7.24)

i) spatial period-doubling occurs in the same manner as in the deterministic case, but

ii) kinks are generated at the upstream, which is due to the phase change of the oscillation at a site where the bifurcation occurs and

iii) the kinks are propagated with a constant speed to the downstream, where the speed is determined by the difference of the phases by a kink while the density of kinks increases as the strength of noise gets larger. The propagation of kinks are shown in Fig. 7.25.

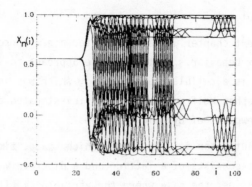

Fig. 7.24 $x_n(i)$'s are plotted for n = 5001, 5002, ... , and
5020 for Eq. (7.5.1) with a noise homogeneously distributed in the interval $(-5 \times 10^{-14}, 5 \times 10^{-14})$.
a = 14 and ε = 0.5, with the same initial condition
as in Fig. 7.21. Note the existence of kinks.

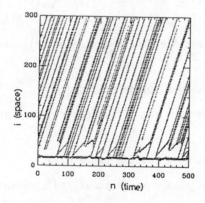

Fig. 7.25 Space-time representation of kink propagations.
The position of kinks are plotted. N = 300,
a = 1.4 and ε = 0.5 and noises are distributed in $(-5 \times 10^{-9}, 5 \times 10^{-9})$. If
$|x_{5000+4n}(i + 1) - x_{5000+4n}(i)|$ exceeds 0.05,
the corresponding space-time position (i, n)
is dotted.

§7.6 Cellular Automata

In the present chapter we have presented a new approach towards spatially complex behavior. Another simple approach to spatially extended systems is a cellular automaton[26], which was introduced by von Neumann. Stephen Wolfram has recently investigated cellular automata from the viewpoint of dynamical systems[27].

A cellular automaton is a system in which space, time, and state are discrete. It consists of a lattice and a discrete variable on a cell. Let us consider the case where the variable $x_n(i)$ (i is a one-dimensional lattice site and n is a discrete time step) takes only two values, i.e., 0 or 1. The dynamics is given by a rule according to which the value of each cell changes per one time step. There are various kinds of rules, e.g., deterministic or probabilistic, nearest neighbor, next nearest neighbor, ···, or long-ranged, irreversible or reversible. Here we consider deterministic, next nearest neighbor, and irreversible one-dimensional cellular automata, i.e., $x_{n+1}(i)$ is determined by $\{x_n(i - 2), x_n(i - 1), x_n(i), x_n(i + 1), x_n(i + 2)\}$.

Wolfram has classified the behavior of cellular automata into four qualitative classes.[27]

(1) Evolution leads to a homogeneous state.

(2) Evolution leads to a set of separated simple stable or periodic structures.

(3) Evolution leads to a chaotic pattern.

(4) Evolution leads to a complex localized structure.

In Figs. 7.26, some examples of cellular automata evolutions are depicted. Wolfram has studied the class-3 cellular automata in detail and clarified the geometry of spatiotemporal patterns, i.e., triangular "clearings" as are seen in Figs. 7.26 d), e). He has calculated various statistical quantities for class-3 cellular automata, such as the distribution of triangles.

Some coupled map lattices, as was shown in §7.4, give quite analogous behaviors to class-3 cellular automata.

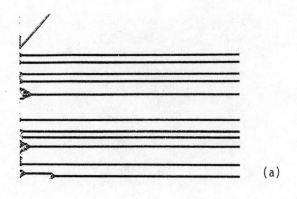

(a)

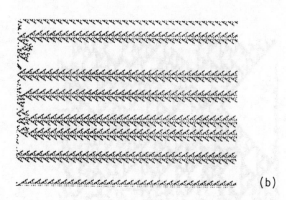

(b)

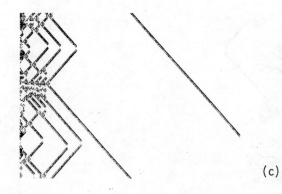

(c)

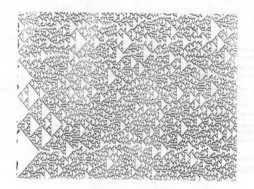

(d)

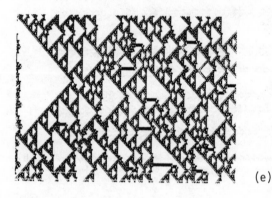

(e)

(f)

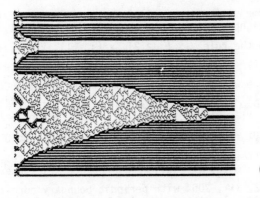

(g)

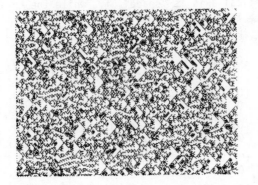

(h)

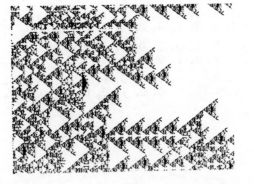

(i)

Fig. 7.26 Some patterns of cellular automata with next nearest neighbor couplings. $x_{n+1}(i)$ takes 0 or 1 according to the rule $\{x_n(i-2), x_n(i-1), x_n(i), x_n(i+1), x_n(i+2)\} \rightarrow x_{n+1}(i)$. A rule is represented by a code which contains 32 numbers (0 or 1). Code $[k_0, k_1, \cdots, k_{30}, k_{31}]$ ($k_j = 0$ or 1) means $\{x_n(i-2) = 0, x_n(i-1) = 0, x_n(i) = 0, x_n(i+1) = 0, x_n(i+2) = 0\} \rightarrow x_{n+1}(i) = k_0$, $\{0, 0, 0, 0, 1\} \rightarrow k_1$, $\cdots$, $\{1, 1, 1, 1, 1\} \rightarrow k_{31}$. In the figures, a dot is plotted if $x_n(i) = 1$. "Space" i takes $i = 1, 2, \cdots, 200$ with periodic boundary condition and "time" n takes $n = 1, 2, \cdots, 300$ initial configuration is random with 50% probability for 1. The vertical axis is "space" i, while the horizontal axis is "time" n. Rules are as follows:

$(k_0, k_1, k_2, \cdots, k_{31}) =$

a) (0, 0, 1, 0, 0, 0, 0, 1, 1, 0, 0, 1, 0, 1, 1, 1,
   0, 1, 0, 0, 0, 1; 1, 0, 0, 0, 1, 1, 1, 0, 1, 0)

b) (0, 0, 1, 0, 1, 0, 0, 1, 1, 0, 0, 1, 0, 1, 0, 0
   0, 0, 0, 0, 0, 0, 1, 0, 0, 0, 1, 0, 1, 0, 0, 0)

c) (0, 0, 1, 0, 0, 0, 0, 0, 1, 0, 0, 1, 0, 1, 0, 1
   0, 1, 0, 1, 0, 0, 1, 0, 0, 1, 1, 1, 0, 0, 1, 1)

d) (0, 0, 1, 0, 0, 0, 0, 0, 1, 0, 0, 1, 0, 1, 0, 0
   0, 0, 0, 0, 0, 0, 1, 1, 0, 0, 1, 1, 0, 1, 0, 0)

e) (0, 0, 1, 0, 0, 0, 0, 1, 1, 1, 0, 1, 0, 1, 0, 1
   0, 0, 1, 1, 0, 1, 1, 1, 0, 1, 1, 1, 1, 1, 1, 0)

f) (0, 0, 1, 0, 1, 0, 0, 1, 1, 0, 0, 1, 0, 1, 0, 1
   0, 0, 0, 1, 0, 0, 1, 0, 0, 1, 1, 0, 1, 0, 1, 0)

g) (0, 0, 1, 0, 0, 0, 0, 1, 1, 0, 0, 1, 0, 1, 1, 1
   0, 0, 0, 1, 0, 0, 1, 1, 0, 1, 1, 0, 1, 1, 1, 0)

h) (0, 0, 1, 0, 0, 0, 0, 0, 1, 0, 0, 1, 0, 1, 1, 1
   0, 1, 0, 1, 0, 1, 0, 0, 1, 1, 0, 0, 1, 1, 0)

i) (0, 0, 1, 0, 1, 0, 0, 1, 1, 0, 0, 1, 0, 1, 0, 0
   0, 1, 0, 0, 0, 0, 1, 0, 0, 0, 1, 1, 1, 0, 0, 1)

The advantage of a cellular automaton over a coupled map lattice is its simplicity. It is easily simulated by a microcomputer. Furthermore, we can calculate a number of attractors rather easily since the state of a cellular automaton is finite if the system size is finite.

A coupled map lattice, on the other hand, has its advantage. It can be related with a partial differential equation system, if the coupled map lattice has a continuum limit. For example, let us take the form

$$x_{n+1}(i) = \sum_{j=-I}^{I} w(j)f(x_n(j + i)) \tag{7.6.1}$$

where I is a range of interactions and $w(j)$ is a weight function which decays slowly as $|j|$ is increased.[28] ($\sum_{j=-I}^{I} w(j) = 1$). Then it is clear that (7.6.1) has a continuum limit.

Another advantage is that we can use various established results on dynamical systems. Especially, recent studies on one-dimensional mappings are utilized if one chooses a familiar one-dimensional map $f(x)$ for the coupled map lattice. Furthermore, the dynamics of an element itself has a complexity in a coupled map lattice, which may be essential for some complex systems.

Both cellular automata and coupled map lattices are simpler than partial differential equation systems. To study the relations among the three systems is an important future problem in nonlinear science.

Class-4 cellular automata are very interesting. Wolfram has conjectured that they have the ability of universal computation in the sense of Furing. An important property of class-4 cellular automata lies in the existence of localized propagating structures, which may be called as "solitons" in a rough sense. Quite recently, Aizawa, Nishikawa and the author have studied 8192 rules of next nearest neighbor cellular automata in which the sequence 1011 can propagate with a velocity= 1.[29] Various types of "soliton" interactions have been found.[29] In some rules, "solitons" can pass through each other with some phase shifts, while they annihilate, or form a self-similar structure of triangles, or form more solitons for some other rules. One interesting aspect is the generation of a "nucleus" (or "soliton gun" in other words), from which solitons are shot periodically. These studies will be important not only for soliton physics in nonintegrable systems, but also for computer science, since such "solitons" can propagate the information.

§7.7 Discussions

In the present chapter we have introduced "coupled map lattice" (CML) to study the field theory of chaos. CML is a new approach towards spatially complex behaviors and many problems are left to the future. In the present section, some of them will be discussed.

(I) Lyapunov spectra

As the bifurcation parameter is changed, CML shows an evolution from simple chaos to high-dimensional hyperchaos. One characteristic quantity for the development of chaos is Lyapunov exponent, as was shown in §7.4. In Fig. 7.27, the number of positive Lyapunov exponents is depicted as a function of the bifurcation parameter A for the transition from torus to chaos from the zigzag structure in §7.3. The number increases roughly as $(A - A_c)^{1/2}$. Similar behavior is obtained also for the case in §7.2.

If the k-th Lyapunov exponent has the form

$$\lambda_k \sim \lambda_0 - ak^2 , \tag{7.7.1}$$

which may be expected from the Fourier analysis and, furthermore, if we assume $\lambda_0 \propto (A - A_c)$, then the above behavior $(A - A_c)^{1/2}$ for the number of positive exponents is derived. From the numerical results on the Lyapunov exponents, the form (7.7.1) seems to be satisfactory.

Detailed study on Lyapunov spectra and also on the fractal dimension of the attractor, however, remains to be a future problem.

(II) Lyapunov vectors and the propagation of disturbance

Lyapunov vector is an eigenvector for the long time product of Jacobi matrices for the mapping, from the eigenvalue of which the corresponding Lyapunov exponent is calculated. Some Lyapunov vectors

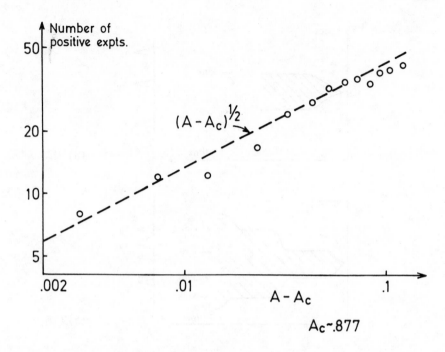

Fig. 7.27 Number of positive Lyapunov exponents as a function
of A for the Map I (see §7.1) with the initial
condition x(i) = sin (2πi/N) and D = 0.2. System
size is N = 50. At the parameters in the figure,
transition from torus to chaos in a zigzag structure
proceeds as is seen in §7.3.

are localized within a finite space, while others are extended. Let us
consider their meanings in connection with the propagation of disturb-
ances.

We apply a pulse at the time step m on the site j and study how
the disturbance propagates. Here we consider two CML systems and compare
the original system $x_n(i)$ with the perturbed system $\bar{x}_n(i)$ for n > m.
Both $x_n(i)$ and $\bar{x}_n(i)$ obey the same CML equation and satisfy $x_n(i)$ =
$\bar{x}_n(i)$ for n < m and $\bar{x}_m(i) = x_m(i)$ for i ≠ j and $\bar{x}_m(j) = x_m(j) + \delta$,
where δ is the disturbance by the pulse. In Figs. 7.28, the spacetime
region where $|\bar{x}_n(i) - x_n(i)| > \delta$ holds is schematically represented.

230

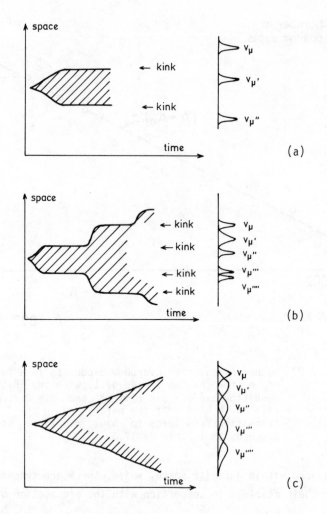

Fig. 7.28 Schematic representation of the propagation of a disturbance. Space-time regions which satisfy $|x_n(i) - \bar{x}_n(i)| > \delta$ are shown with Lyapunov vectors v_μ for positive exponents. In a), chaos and v_μ are localized in the region between kinks. In b), tunneling occurs due to the overlapping of Lyapunov vectors. In c), the propagation is rather smooth. These behaviors are found from numerical results for the coupled map lattices at the parameters where band-mergings occur for the logistic map (see §7.2).

If the difference $\Delta x_n(i)$ between $\bar{x}_n(i)$ and $x_n(i)$ remains small, it is represented by Lyapunov analysis as

$$\Delta x_n(i) = \sum_\mu v_\mu(i) e^{\lambda_\mu(n-m)} \Delta_\mu$$

where v_μ and λ_μ are Lyapunov vectors and exponents and $\Delta x_m(i) = \sum_\mu \Delta_\mu v_\mu(i)$.

In the case for Fig. 7.28a), the disturbance is localized within a limited space. One example for this case is the pattern in Fig. 7.4, where kinks are stable and fixed and the disturbance cannot propagate across the kinks. In this case, the Lyapunov vectors corresponding to positive exponents are localized in space.

Furthermore, the vectors $v_\mu(i)$'s for positive exponents which are localized near $i = j$ do not have any overlappings (or too little if they have any) with $v_\kappa(i)$'s which are localized in the region beyond the kink.

If the vectors have little overlappings, the disturbance can propagate after some waiting time. In this case, a kind of "tunneling" phenomenon occurs for the propagation, as is shown in Fig. 7.28b).

If the vectors have sufficient overlappings or the vectors are extended for positive exponents, the disturbance can propagate all through the space, as is schematically shown in Fig. 7.28c).

For the model in §7.2, the transition from the pattern in Fig. 7.28a) to b) and then to c) has been observed as A is increased.

Detailed study, again, is left to the future.

(III) Direct product state of chaos

As is suggested in §7.4, it may be possible to construct a field theory of chaos on the basis of "mean field theory" and direct product state assumption. In a variety of systems, this assumption seems to be approximately true though the detailed study is left to the future.

(IV) Open flow systems

In §7.5, some results for the CML with one-way coupling are presented. Studies of open flow systems from the viewpoint of dynamical systems theory have just been started and there remain a lot of future problems, such as the change of dimension as the move of the flow to downstream, the propagation of disturbance, and the statistical property of turbulence at the downstream.

Up to now, it is not sure whether the spatial period-doubling can be observed in the open flow experiments or in the numerical simulation of the partial differential equation in the presence of the first spatial derivative. It will be of interest to make such experiments in open fluid systems as to take a Poincaré map or to measure the power spectra at various points of the flow.

REFERENCES

*) For a part of the contents of the present chapter see K. Kaneko, Prog. Theor. Phys. 72 (1984) 480; pp 272-277 in Dynamical Problems in Soliton Systems (Springer, 1985; ed. S. Takeno); "Turbulence in Coupled Map Lattices", to appear in Physica D; "Spatiotemporal Intermittency in CML" Prog. Theor. Phys. 75 (1985); Phys. Lett. 111A (1985) 321.

1. D. Ruelle and F. Takens Comm. Math. Phys. 20 (1971) 167.
   S. Newhouse, D. Ruelle and F. Takens, Comm. Math. Phys. 64 (1978) 35.

2. Turbulence and Chaotic Phenomena in Fluids (North Holland, 1984; ed. T. Tatsumi).

3. Y. Kuramoto, Physica 106A (1981) 128.

4. K. Ikeda, H. Daido and O. Akimoto, Phys. Rev. Lett. 45 (1980) 709.

5. J.C. Eilbeck, P.S. Lomdahl and A.C. Newell, Phys. Lett. 87A (1981) 1.

6. D. Bennet, A.R. Bishop and S.E. Trullinger, Z. Phys. B47 (1982) 265.

7. See e.g., J.W. Clark, J. Rafelski, and J.V. Winston, Phys. Rep. 123 (1985) 216.

8. J.D. Farmer, Physica 4D (1982) 366.

9. Y. Pomeau, A. Pumir and P. Pelce, J. Stat. Phys. 37 (1984) 39.

10. Dynamical Problems in Soliton Systems (Springer, 1985; ed. S. Takeno).

11. See for recent advances in CMLs, R.J. Deissler, Phys. Lett. 100A (1984) 451; J. Crutchfield, private communication; J.D. Keeler and J.D. Farmer, private communication; R. Kapral, Phys. Rev. A31 (1985) 3868; Y, Aizawa, Prog. Theor. Phys. 72 (1984) 662; Y. Aizawa and I. Nishikawa, private communication; T. Yamada and H. Fujisaka, Prog. Theor. Phys. 72 (1985) 885.

12. W.P. Su, J.R. Schrieffer and A.J. Heeger, Phys. Rev. Lett. 42 (1979) 1698.

13. Y. Pomeau and P. Manneville, Comm. Math. Phys. 74 (1980) 189.

14. E.A. Novikov and P.W. Stewart, Isv. Akad. Nauk. USSR, Ser. Geophys. 3 (1964) 408; B.B. Mandelbrot, J. Fluid Mech. 62 (1974) 331; U. Frisch, P.L. Sulem and M. Nelkin, J. Fluid Mech. 87 (1978) 719.

15. See Chap. 3.

16. S. Takesue and K. Kaneko, Prog. Theor. Phys. 71 (1984) 35.

17. C. Grebogi, E. Ott, and J.A. Yorke, Physica 7D (1983) 181.

18. D.K. Campbell, J.F. Schonfeld and C.A. Wingate, Physica 9D (1983) 1.

19. S. Wolfram, Rev. Mod. Phys. 55 (1983) 601.

20. I. Shimada and T. Nagashima, Prog. Theor. Phys. 61 (1979) 1605.

21. K. Sreenivasan, in Frontiers in Fluid Mechanics (eds. S.H. Davis and J.L. Lumley; Springer, 1985).

22. See e.g., R. Feynman, R.B. Leghton and M. Sands, Feynman Lectures on Physics Vol. 2 (Addison Wesley Pub. Co., 1964).

23. M.J. Feigenbaum, J. Stat. Phys. 19 (1978) 25, 21 (1979) 669.

24. R.J. Deissler, J. Stat. Phys. 40 (1985) 371.

25. R.J. Deissler and K. Kaneko, preprint (Los Alamos, 1985).

26. See for Cellular Automata, Physica 10D (1984); eds. D. Farmer, T. Toffoli, and S. Wolfram.

234

27. Ref. 19) and S. Wolfram in ref. 26).

28. J.D. Keeler and J.D. Farmer, private communication.

29. Y. Aizawa, I. Nishikawa and K. Kaneko, to be published.

Chapter 8

SUMMARY, FUTURE PROBLEMS, AND DISCUSSIONS

南海之帝爲儵、北海之帝爲忽、

中央之帝爲渾沌、儵與忽、時

相與遇於渾沌之地、渾沌待之

甚善、儵與忽、謀報渾沌之德、

曰、人皆有七竅、以視聽食息、

此獨无有、嘗試鑿之、曰鑿一

竅、七日而渾沌死

莊子・卷之七

§8.1 Summary and Future Problems

In the present book we have studied various aspects on the collapse of tori. Phase instability of a torus has been investigated in Chap. 2, where locking structures in the circle map (2.1.2) are studied in detail. Amplitude instability of a torus motion is investigated in Chaps. 4 and 5, where oscillation, fractalization and doubling of tori are found and analyzed. Stability of a 3-torus is confirmed in Chap. 6. In Chap. 3, we have studied a coupled logistic map which is the starting point of a coupled map lattice study, which is a new and useful method for the spatially complex system. In Chap. 7, some results on coupled map lattices are shown, such as period-doublings of kink-antikink patterns, transition from torus to chaos in a zigzag structure, geometrical structures in spatiotemporal intermittency, and spatial period-doubling in open flow.

We have used and introduced verious mappings to study the collapse of tori. The circle map (2.1.2) is now regarded as an important and typical model for the study of phase dynamics. We have also used "coupled maps", "delayed maps", and "modulation maps". They have important meanings by themselves and will be useful to study coupled systems, delay-differential equations, and systems with (incommensurate) periodic modulation respectively.

Studies of the collapse of tori in dissipative systems have just started. Though many important works have appeared these few years, there are a lot of problems left to the future. Here, we will list the future problems which are related to the present book.

a)   Instability of phase dynamics of the map (2.1.2).

  a1) Global structure of a phase diagram for $A > 1/(2\pi)$.

  a2) Characterization of chaotic orbits, especially in connection with the disordering property. Quantitative argument of the disordering property. Relation between the disordering and other quantities such as power spectra.

a3) Symbolic dynamics approach to the disordering property.

a4) Study of the multibasin phenomena for $A > 1/(2\pi)$. Stability of attractors which coexist for given parameters A and D. Basin structure.

a5) Further studies on the similarity structure of the phase diagram. Not only the similarities of window structures but also the self-similarity in chaos-chaos transitions[1] (e.g., band splitting or crisis).

a6) Noise effect. Small structures of a devil's staircase disappear successively as the strength of a noise is increased. How small structures can be observed for a given noise? Are there any scaling relations between the strength of a noise and some properties of a locking (e.g., a period)? Noise effect on a chaotic orbit is also important. Does noise decrease or increase the chaotic properties, such as the Lyapunov exponent, entropy, and disordering ratio?[2] Does the disordering time distribution change in the presence of noise?

a7) Detailed study on the global property of lockings and windows. Relation between the logistic map and the circle map (see §2.8).

b) In connection with a period-adding sequence.

b1) Is it possible to construct a renormalization group theory which gives the fixed point function?

b2) Can our theory be of relevance to study other types of period-adding sequences such as the one observed in B-Z reaction? (See also §2.6.)

b3) Extension of our theory to the period-adding sequences in other classes, such as the ones corresponding to the type II or type-III intermittency (the first step towards this extension is already given in §2.6) and also, extension to area-preserving mappings, such as a standard mapping.

c) Transition from torus to chaos in two-dimensional mappings.

    c1) Development of chaos after the collapse of tori. Quantitative studies on the increase of the dimension of the attractor and the increase of the width of the belt-like attractor (see also §3.5 and §4.4). Mechanism of a chaos-hyperchaos transition.

    c2) Supercritical behavior at the collapse of tori. Can one find any critical behavior for the Lyapunov exponent, the fractal dimension, and the integrated noise of power spectra etc? (Do they show the behavior like $(a - a_c^{\alpha})$?)

    c3) It is of relevance to study the two-dimensional map

$$\theta_{n+1} = \theta_n + D + b(r_n - R) + A \sin (2\pi\theta_n)$$

$$r_{n+1} = R + b(r_n - R) + A \sin (2\pi\theta_n) \quad . \tag{8.1.1}$$

The relation between the map (8.1.1) and the circle map (2.1.2) corresponds to the relation between the Henon map (1.2.4) and the logistic map (1.2.3). As the dissipation b^{-1} goes to infinity, the map (8.1.1) reduces to the circle map. For $D = R$ and $b = 1$, we have the standard map (2.10.1). There are many problems on the map (8.1.1), such as the relation between the dimension and the dissipation, multibasin phenomena, the relation between the strength of dissipation and the number of basins, and the mechanism of the development of chaos.

d) In connection with coupled logistic maps.

    d1) Mechanism of symmetry breaking. Significance of the oscillation with broken symmetry as a model for the spatio-temporal structure in nonlinear nonequilibrium systems.

    d2) Mechanism of chaos-chaos transition (such as the fusions of chaos in Figs. 3.2).

e) Delayed maps.

e1) A delayed map is introduced by a discretization of a
difference-differential equation. As the number of mesh points
N for a discretization goes to infinity, the map approaches
the original equation. Is there any relation between the
number of the observable doubling and the mesh points N?
Since the original equation corresponds to an infinite-
dimensional mapping, a "turbulent" state can exist (see also
Ref. 3). It will be important to study this type of turbulence
(see also §8.3).

e2) The delayed piecewise-linear map (4.2.2) is the simplest and
nontrivial model among delayed maps. Though it is simple, it
shows a very interesting attractor (Fig. 4.7). We can explain
qualitatively the mechanism of the oscillation, but we do not
have any quantitative results. Since the map is piecewise-
linear, it will be possible to obtain analytic results on the
Lyapunov exponent, width of the attractor, wavelength and decay
rate of the oscillation, and critical phenomena as
$D \to 1/(1 - A) + 0$ etc.

f) Fractalization of tori.

f1) Detailed mechanism of the fractalization of tori. We can say
something about the mechanism, by using the continued fraction
approximation $C \approx F_{k-1}/F_k$ (see §4.3) and consider the one-
dimensional map $f^{F_k}(x)$. The detailed study on the mechanism,
however, remains as a future problem.

f2) Further numerical calculations of the fractal dimension at the
onset of chaos. Especially, it will be of importance to study
its dependence on rotation numbers (or on the tails of their
continued fraction expansion).

f3) To confirm the universality.

f4) Detailed study on the critical phenomena. Is the finite size
scaling approach possible? Is it possible to find other

critical exponents (such as the crossover exponent)? Of course, it is important to construct a renormalization group theory.

f5) Supercritical behavior. Are there any critical behavior for the Lyapunov exponent and the width of the belt-like attractor etc. (e.g., the behavior $(\varepsilon - \varepsilon_c)^{\alpha}$) for $\varepsilon > \varepsilon_c$? How does the chaos develop for $\varepsilon > \varepsilon_c$? How is the chaotic orbit characterized?

f6) In the models in §4.3, the tori with the rotation number $C = (\sqrt{5} - 1)/2$ collapses at a smaller value of a coupling ε than the tori with $C = (\sqrt{2} - 1)$. Then, which torus (i.e., the torus with what rotation number) is the last or first to collapse? Is it possible to construct a theory like the KAM theory in conservative systems?

f7) The functional map in §4.3 may be an interesting model to consider a spatially complicated pattern. (This problem is discussed in §8.3.)

g) Doubling of torus.

g1) Detailed mechanism of the interruption of the doubling cascade.

g2) Is the functional map approach in §4.3 relevant to study the doubling of tori?

h) Stability of a three-torus.

h1) Is it possible to find a simple model which shows Ruelle and Takens' scenario? Can the exponential noise in Libchaber's experiment be explained?

h2) Is it possible to construct a theory for a three-torus (or a four-torus etc.) in dissipative systems like the KAM theory in conservative systems?

h3) Can one say anything quantitative about the regions of lockings in a three-torus?

h4) Does an n-torus $(n \geq 4)$ exist stably? The regions of lockings will be expected to increase as n increases. Is it possible to obtain a quantitative result for this expectation?

h5) Study of the similarity structures and scaling properties in "double devil's staircases". Is it possible to extend a continued fraction expansion method to the problem of double (or triple etc.) devil's staircase?

h6) Development of chaos after the collapse of tori, especially the increase of the dimension.

i) Stability of a coupled system.

i1) In the present thesis, we have used various coupled maps to study the stability of direct product state, such as cycle $\otimes$ cycle, cycle$\otimes$torus, torus$\otimes$torus, cycle$\otimes$chaos, torus$\otimes$chaos, and chaos$\otimes$chaos. Among these states, the direct product state "chaos$\otimes$chaos" seems to be most stable against a structural perturbation. Can one say anything theoretical about this observation? (Pseudo orbit tracing property of a chaotic orbit may be a key to study this phenomenon).

i2) Stability of a bifurcation. Coupling a period-doubling bifurcation with a torus, we have found that the doubling cascade does not continue infinitely (§5.2). Coupling two period-doubling bifurcations, we have again found that the doubling occurs only a finite number of times.[4] Thus, an infinite cascade of bifurcations seems to be unstable against a coupling. It may be of interest to verify the above conjecture for other types of bifurcations, such as a Hopf bifurcation and to search for a relation between the number of bifurcations and a coupling.

(For the problems in Chap. 7, discussions will be shown in §8.3.)

Our method is confined within a mapping. History of nonlinear physics research has told us that the interesting phenomena which have been found in "usual" mappings also appear in a flow system with higher-dimensions. Thus, it will be of interest and importance to verify that the new phenomena in the present book can also exist in flow systems.

Of course, there are many important future problems in addition to the above list, even if our interest is confined to the problems which are directly related to the transition from torus to chaos.

Another topic which is not treated in the present book is torus intermittency[5),6)]. The amplitude motion of a torus shows the instability of Pomeau and Manneville's types[7)] in some cases. Daido[5)] has investigated various classes of torus intermittency recently. Experimentally, some examples have been observed in the Bénard convection[8)].

If we consider the general aspects of chaotic phenomena, there are much more important and interesting problems left to the future, which are fundamental to our understanding of nature. Some examples are discussed in the following two sections.

§8.2 What has Chaos Brought about and will Bring about in Science?

a) Collapse of Determinism and Observability

Our science consists of two parts, i.e., predictions from laws and discoveries of laws from data. As is well-known, the discovery of chaos has brought about the limitation to the predictions. Does the chaos also give some limitations to discoveries of laws?

Recently, algorithms have been proposed to calculate a fractal dimension[9),10)] and to construct a finite-dimensional dynamical system[11)] from data (the dimension is a little larger than the fractal dimension). Thus, one might conclude that it has become possible, in principle, to obtain a law from data. There seems to exist, however, an essential limitation to the above algorithms. As the fractal dimension becomes larger or the nonlinearity gets stronger, the complexity[12),13)] of the data becomes larger. It is proven, however, that it <u>cannot be proven</u> that a set of data has a complexity larger than $n + c$, in a formal system with the complexity n (c is a constant)[12)]. Since our formal system (e.g., mathematics or a computer system) has a finite complexity, it is impossible to extract a law from a set of data with very large complexity or to conclude that the data are random. Thus, there seem to exist essential limitations to our knowledge. We cannot conclude that there is no law (i.e., random).

It will be of importance to search for the relations between the limit of our knowledge for a given formal system and the accuracy of the observation, the entropy or dimension of the observed data. It will also be important to construct an algorithm to extract a law and some fluctuations (noises) from a set of data and to check its limitation. Noise-induced order[2] may play an essential role for this limitation.

Finiteness of our computer systems gives a limitation to numerical studies of chaos.[13] In our finite-state machines, any orbit falls on a cycle ultimately. Thus it will be necessary to search for the relationship between the physical chaotic behavior and the behavior observed in a computer. To make clear the above round-off problem, we consider a model

$$x_{n+1} = \frac{1}{N} [f(x_n) \times N] \tag{8.2.1}$$

for initial conditions $x_0 \in \{i/N\}$ (i = integer) ([y] denotes the integer part of y). We assume that the orbits are confined in a finite domain. Then the problem reduces to the following deterministic application

$$x_n^i (= \frac{i}{N}) \to x_{n+1}^j \quad (-Nc' \le i,j \le Nc) \tag{8.2.2}$$

where c and c' are finite constants. As the number of mesh points N is increased, the accuracy of "our computer" increases. If the true orbit for the map $x_{n+1} = f(x_n)$ is chaotic, the period for a cycle is expected to go to infinity as $N \to \infty$. It will be of importance to study how the following quantities increase as N is increased;

  i) numbers of ultimate cycles,

 ii) periods for cycles and

iii) length of transients before an orbit falls on the ultimate cycle.

In Fig. 8.1, a period for a cycle with the initial conditions

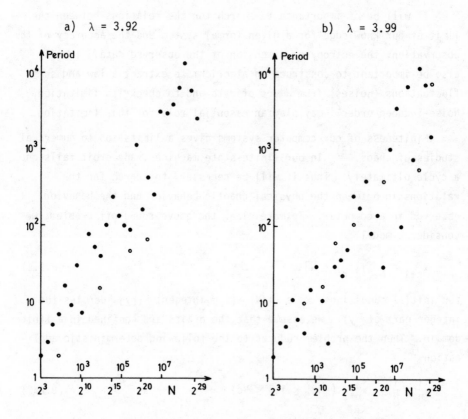

Fig. 8.1   Period of the map (8.2.1) as a function of the
number of mesh points   N,   where   f(x) =
λx(1 - x).   Initial value of   x   is chosen to
be   x = 0.5   (•)   and   x = 0.25   (o;   plotted
only if the period differs from the value for
x = 0.5).   The mesh is changed by taking the
values   N = 2^k   (k = 3,4, ⋯, 29).

$[N/k]/N$ $(k = 2,4)$ is plotted for the logistic map $f(x) = \lambda x(1 - x)$
as a function of N. Detailed study remains as a future problem. It
will also be interesting to investigate the mechanism of the increase of
a period (doubling etc.).

b) Encounter with Infinity

  i) Self-Similarity

    It is one of the most marvellous achievement of our mind that the
notion "infinity" has been introduced to mathematics and sciences, though
our brain consists of a finite number of neurons. Up to now, our understanding of infinities is essentially based on the recursion relations.
The most typical example of these recursion relations is a renormalization
group, which is the most powerful method to treat a self-similar structure. Renormalization group was invented to treat the "infinity" in the
field theory, i.e., the ultraviolet divergence. Since the author is not
an expert in the field theory, this problem will not be referred in the
following.

    Statistical physics usually involves an infinity. In many cases,
however, this infinity is rather trivial, because Gaussian distribution
is dominant owing to the central limit theorem and we are not involved
in difficult problems about the infinity. Non-Gaussian distribution was
mainly introduced to statistical physics by a theory of phase transitions.
Renormalization group (RG) approach has made a great success in the study
of phase transitions, which was introduced by L.P. Kadanoff and K.G.
Wilson[14].

    Onset of chaos resembles phase transitions in equilibrium statis-
tical physics in various aspects, which recent success of RG approach has
clearly shown.[15] The success of RG is based on the self-similarity at
the critical point. The self-similarity is excellently expressed by the
fixed point function of a RG transformation. Thus, physicists have
succeeded in treating a "nontrivial" infinity.

    However, this success must not be overevaluated. In these problems
(i.e., phase transitions and onset of chaos), nontrivial infinity appears

only at one point, i.e., at one critical point[16] (at the transition temperature $T = T_c$ or at a parameter of the onset of chaos).

In chaos, however, we always encounter a nontrivial infinity (not only at one critical point), as can be remarkably seen in Fig. 1.5 (Hénon attractor). Enlargement of Fig. 1.5 is repeated infinite times in principle, which gives a similar figure. A strange attractor is fractal in this sense, as was pointed out by B.B. Mandelbrot.[9] Recently, many attempts to obtain the fractal dimension numerically or experimentally have been performed.[10]

How does the self-similarity reflect on the properties of time series, such as the autocorrelation function and the power spectra? It is an important question, but the answer has not yet been obtained (so far as the author knows). Renormalization group approach might be useful to study the self-similarity in chaos itself, which remains as a future problem.

In order to see smaller structure of the Cantor set, longer time for the observation or computation is necessary. Thus, infinite time is necessary to get the complete information for the strange attractor. In this sense, we have encountered the problem of "non-stationarity" in chaos, which will be discussed in the next subsection.

Another important problem is a fine structure in a parameter space. For example, chaos cannot exist in an open interval in the parameter λ for the logistic map $x_{n+1} = \lambda x_n(1 - x_n)$. The measure of chaos in the parameter space, however, is not zero. For the circle map and coupled circle maps, situations are quite similar. Tori, three-tori and chaos cannot exist in an open set, but have finite measures. Thus, we have to "live with structural instability"[17]. Quantitative studies on the self-similar structure in the parameter space are rather few, but they have to be done in future, since we have to abandon the determinism in the usual sense in a structural unstable system and look for some quantities which characterize the limit of observation (the limit of determinism).

ii) Non-Stationarity

In linear nonequilibrium statistical physics, we have already encountered with long-time behavior. Alder and Wainwright have found a long-time-tail in fluids,[18] which cannot be explained by a simple study by the Boltzmann equation approach, where the distribution function approaches equilibrium exponentially. Critical slowing down at the phase transition point[19] is also an example of a long-time-tail. The f^{-1} spectrum in a variety of systems[20] gives another example of the long time behavior.

Recently, Aizawa et al.[21] have investigated the intermittent chaos and found f^{-1} spectra (see also §1.3). They have used symbolic dynamics approach to extract the Paretian distribution, which is typical in the social science.[22],[23]

Though their work is important to understand the long-time behavior at the onset of chaos via intermittency, a non-stationary feature in chaotic orbit itself, which is described at the end of i), has not yet been understood.

Non-stationarity in chaos in a one-dimensional map will also be seen in the singularities in the invariant measure. The invariant measure for the logistic map $x_{n+1} = \lambda x_n (1 - x_n)$ has singularities $(x - x_i)^{-1/2}$ at $x = x_i$ ((x_i) are successive image points of 0.5 by the map) for the parameter value λ where chaos exists. A very long time is necessary to obtain the invariant measure accurately. Thus, the singularity in the measure (which is a typical example of a non-Gaussian distribution) may give the non-stationarity in the chaos. It is also interesting to search for the relationship between the Cantor structure in the Hénon map (Fig. 1.5) and the singularities in the measure for the logistic map.

Another possibility for nonstationarity is the stochasticity in Hamiltonian systems. For example, self-similar structures of islands in stochastic orbits for the standard map (2.10.1) have recently been found numerically.[24],[25] Such hierarchy of islands seem to bring about f^{-1} spectra.[26] In higher-dimensional systems, Arnold diffusion makes a stochastic trajectory to explore the whole energy surface.[27] Arnold diffusion, however, is very slow and an orbit is trapped near a KAM torus,[28] which seems to cause a kind of nonstationarity. Thus, an f^{-1} type of spectra might be generic in high-dimensional Hamiltonian systems.

To sum up 7.2b) i) and ii), we have many important future problems on the "nontrivial infinity" in chaos itself.

iii) Irrational Physics

Quasiperiodic states involve a kind of infinity, though it seems simpler than the infinity for the chaos. The infinity is due to an irrational number (we need infinite digits for an irrational number). Up to recent times, it has hardly been necessary for physicists to distinguish whether some quantities are irrational or not. Recent studies on commensurate-incommensurate transitions, localization in a quasi-periodic potential, and stability of a torus may open a new field of irrational physics, where some techniques in the present book (such as continued fraction expansions) may be powerful (see also §2.10).

c) Impact on Various Fields

Discovery of chaos in dissipative systems and the observation that the chaos is a common and usual behavior in nonlinear systems has made and will make a large impact on various branches in science. In this subsection we briefly discuss future problems on the impact.

i) Biology

Recently chaos has been observed in various examples, such as the rhythm of cardiac cells and neurons[29]. In these cases, however, biological significance of chaos is not clear. On the contrary, experimental situations seem to be out of usual physiological conditions, or chaos seems to correspond to "illness".[30]

Does chaos always appear as a harm to biological systems? Perhaps, the answer is no. There can be a lot of positive roles of chaos in biological systems.[31],[32]

First, the soft response of a chaotic orbit against an external perturbation will be important as was pointed out by K. Tomita[31].

Secondly, redundancy in a chaotic orbit may be useful to living things.[32)] The notion of redundancy was introduced by an information theory[33)] and has been applied to linguistics and music etc. Let us consider the music[34),35)]. If the music obeys too strict rules, the redundancy is large and the information gain per one musical note is small. On the other hand, if the rules are too loose, the information gain is large, but the speed of the gain may be faster than our channel capacity. Furthermore, one cannot understand a large part of the music if one fails to hear one musical note, since the redundancy is very small. Thus, the music with very little rules is a nuisance to us.

Let us consider the artificial music[36)]. We take a time series of a variable $\{x_n\} \in [0,1]$ and assign a musical note by $0 \le x < 1/7 \leftrightarrow$ mi, $1/7 \le x < 2/7 \leftrightarrow$ fa, $\cdots$, $6/7 \le x < 1 \leftrightarrow$ re. If $\{x_n\}$ is periodic, the rule is so strict that almost no information is gained and the music is boring. If $\{x_n\}$ is constructed by a random number (no rule), the music is a nuisance. On the other hand, if $\{x_n\}$ is chaotic, information gain exists for every new musical note. Furthermore, we can miss some notes, because there is a strong correlation between nearby notes (redundancy). By listening to the periodic, random, and chaotic music, you may agree that the redundancy in a chaotic orbit might be relevant to living things. (See Fig. 8.2a), b), c) for sheet musics for "torus",

a) "torus music" — $\{x_n\}$ is a time series of the map
with $A = 0.1$ and $D = 0.6$.

b) "chaos music" — $\{x_n\}$ is a time series of the map (2.1.2)
with $A = 0.254$ and $D = 0.6$

c) "random music" — $\{x_n\}$ is a random number which takes
the value homogeneously in [0,1].

Fig. 8.2 Music Sheets for torus, chaos and random.

"chaos" and "random", respectively. It is a little difficult to distinguish chaos from torus for a short sheet music. For a long sheet, however, we can distinguish them rather easily. The torus music gives us the feeling of rotation.)

It will also be important to study how the efficiency increases (decreases) when the biochemical reaction shows a chaotic behavior.

ii) Information Science

Information science was developed by Shannon[37]. Information is gained when one measures something about a state and increases the knowledge about the state. If we can assign probabilities p_i to each possible outcome i, then the information associated with the outcome is defined as

$$H = - \sum_i p_i \ln p_i \qquad (8.2.3)$$

according to Shannon. Here we will discuss informational aspects of chaos, following Robert Shaw's thesis[38].

Let us assume that we are measuring some variables, the change of which is governed by a dynamical system. The information gain is given by the logarithm of the ratio of states distinguishable before (Ω_i) and after (Ω_f) some interval:

$$\Delta H = \log (\Omega_f/\Omega_i) \ . \qquad (8.2.4)$$

If the attractor is periodic or quasiperiodic, the number of distinguishable states cannot increase faster than the power law $\Omega(t) \sim t^n$. Thus, the information creation rate dH/dt of such a system is not positive as time passes.

If the number of distinguishable states increase $\Omega(t) \sim e^{\lambda t}$ ($\lambda > 0$), the system can be an information source. As a simple case, let us consider the one-dimensional map

$$x_{n+1} = f(x_n) = 2x_n \quad (\text{mod } 1) \qquad (8.2.5)$$

If we observe whether x_n is greater than 0.5 or not in each time step, more and more information is obtained as to the initial value x_0. The information gain per iteration is computed from the slope (df/dx) of the map. In the case (8.2.5), the slope is everywhere two and the information gain per iteration is

$$H = \log (2/1) = 1 \text{ bit} .$$

As is easily expected, the information gain is given by the Lyapunov exponent in one-dimensional mappings which show chaotic behavior.

The information gain in chaos is also formulated in flow systems, in connection with Lyapunov exponents. Information dimension was investigated by J.D. Farmer[3], which is an important quantity to characterize chaos.

Shaw has also discussed the informational aspects of turbulence. In turbulence, information flows from microscopic to macroscopic scales. According to Shaw's estimate[38], about 10^{12} bits per second of information is generated by a cup of coffee stirred, which you might want to take while you are reading the present chapter and feel sleepy.

iii) Artificial Intelligence, Brain, and Chaos (ABC)

In the famous science fiction "Solaris"[39], Lem has assumed the existence of the sea with intelligence. Perhaps, this is the first suggestion that turbulence might have an intelligence. In the fiction, the intelligence of the sea cannot be comprehensible by the human intelligence. On the contrary, some people have recently been discussing the possibility that chaos or turbulence is made use of in the human brain or is applied to construct an artificial intelligence. Here we will breifly discuss some aspects of dynamical systems necessary for the study of intelligence.

(a) Creation of information — this is an important property for the intelligence. As has already been discussed in iii), chaotic systems have this ability. The idea of symbolic dynamics approach is essentially related with the theory of generative grammar by Chomsky[40]. The richness of a symbolic sequence in a chaotic orbit may be of relevance for a language theory.

(b) Storage of information — memory is essential both for the brain
and computer. Dynamical systems can be used as a memory in the
following two ways.

One way is to preserve the information in a time series. This
is possible if the system has a chaotic orbit. Nicolis and
Tsuda[41] have discussed the possibility of the existence of the
above mechanism in the human brain, in connection with the magic
number 7 ± 2 problem in psychology.[42] In this case, memory is
stored only for a limited time interval and is lost after some time.

Another way is more naive. If a dynamical system has a large
number of attractors, information is stored in each attractor.
Hopfield has used a kind of cellular automaton as a content address-
able memory,[43] which is related with a spin glass model. Cellular
automata and coupled map lattices can have a large number of attractors, which might be useful for the storage of information. The
notion of attractors is also important as the problem of stability
or self-repair.

(c) Propagation of information — Information processing is necessary
for intelligence. For this purpose, a dynamical system with a
large number of periodic attractors is not suitable, since the
information cannot propagate among local structures. As for this
problem, coupled map lattices with a large number of chaotic
attractors or class-4 cellular automata are useful as can be seen
in Chap. 7.

(d) Hierarchical structure — Our notions constitute a hierarchical
structure in the brain. In computer systems also, tree-like structures are useful as is seen in a UNIX system.[44] Strange attractors have a self-similar structure. It seems, however, rather hard
to make use of the structure as the origin of hierarchy. Another
possibility is to utilize the geometrical structure in space-time
in coupled map lattices or cellular automata, (see Chap. 7) which
has not yet been carried out.

One short cut is to start from the hierarchical structure.

One celebrated example is a (neo)cognitron[45], which has recently
been investigated by d'Humieres and Huberman[46] from the viewpoint
of a dynamical system theory. Coupled map lattices with one-way
coupling or unidirectional cellular automata[47] may be suitable
for this purpose.

(e) Interaction with the outer world — Information must be injected
from outside and be extracted from some points. This easy-looking
imposition is rather hard for simple dynamical systems. Again,
coupled map lattices or cellular automata with one-way coupling may
be useful for this purpose.

Response against an external input is important. As has
already been described in i), Tomita has pointed out that soft
response in chaotic systems is relevant for the pattern recognition
or learning in a brain. Nagumo and Sato[48] have studied an artificial neuron model by a piecewise-linear mapping. In the model,
the output forms a "devil's staircase" as a function of an input.

(f) Adaptation and evolution — One of the marvellous aspects of an
intelligence is that it evolves by adapting itself to the environ-
ment. The adapting feature is important not only in the brain but
also in the immune system and (also of course) in Darwinian evolution. In the evolution of species, for example, new species
appears as time passes, while couplings among the neurons change
in time. Thus, these problems may involve a kind of non-
stationarity. The nonstationarity in chaos seems to be of relevance for this aspect. These biological systems, however, cannot
be represented as a closed set of dynamical systems since the
number of variables or the parameters themselves change in time.
Thus, we have to go beyond the usual dynamical systems. One-way
coupling models may be of relevance since the system evolves as it
goes downflow. Studies of adaptive cellular automata or coupled
map lattices will be essential, where the coupling is changed as
time in reference to the input.

To sum up, dynamical systems with complexity are useful for
artificial intelligence studies or brain, but it is desirable to go

beyond the usual dynamical system in future.

iv) Social Science

In 1963, B.B. Mandelbrot wrote a remarkable paper,[22] where he pointed out the importance of Paretian distributions in social science. He showed that a Paretian distribution is invariant against weighted mixture, maximizing choice, and aggregation. This invariance may be a reason for the relevance of Paretian distributions in social science, but the origin of one Paretian distribution remains unknown. Chaos, if it exists in the dynamics in social science, may be a possible cause to the Paretian distribution. Mandelbrot also pointed out that the non-stationary nature in Paretian distributions made the study of social science very difficult.

Difficulty in social science may also lie in the limitation to the algorithm "data → law" (see §8.2a)). Fractal dimension of the data in social science may be so large that it is impossible for us to conclude the existence of laws. It will be of interest to carry out the algorithm and search for a law for some data in social science.

To sum up, study of chaos may open a new perspective towards social science.

v) Orthodox physics

It goes without saying that chaos physics has made a large impact on atomic physics, nuclear physics, solid-state physics, cosmology, optics, and so on. In solid-state physics, we can see some examples of applications of chaos in the following problems; irrational physics (see §2.10 and §8.2b) iii)), some suggestions about the relation between glassy state and a chaotic behavior[49],[50], possible meanings of chaotic renormalization group,[51] chaotic response against an external field in nonlinear systems such as Josephson junction[52], Suhl instability[53] and so on. In nonlinear optics Ikeda[54] pointed out a new instability which leads to chaos. It has recently been observed experimentally.[55]

Quantum chaos in Hamiltonian systems is a very important problem. Distribution of energy levels in chaotic systems has been numerically obtained in a variety of systems.[56]   M.C. Gutzwiller[57] uses a semi-classical quantization based on path integral formulations, to obtain the energy levels in anistropic Kepler problems. Development of quantum chaos is essential to atomic and nuclear physics.

In fluid mechanics and in field theory, study of chaos in high-dimensional systems is necessary. Field theory of chaos is important for this purpose, which will be discussed in the next section.

§8.3 Towards a Field Theory of Chaos

Turbulence is a rather common phenomenon in nonlinear nonequilibrium systems. Though the distinction between the turbulence and the chaos is not clear, the former usually includes a larger number of modes. Examples of the turbulence (in a wide sense) are fluid turbulence, chemical turbulence,[58] turbulence in a difference-differential equation,[3],[54],[59],[60] (optical turbulence[54] etc.) and a damped sine-Gordon equation with an external perturbation,[61] etc. These systems obey nonlinear partial differential equations or infinite-dimensional mappings. It is generally considered that an infinite number of modes are excited (i.e., there are an infinite number of positive Lyapunov exponents) in the fully-developed turbulence (e.g., fluid turbulence with Reynolds number $\to \infty$).

Why is a theory of turbulence difficult? In a field theory and equilibrium statistical physics, we have succeeded in treating the problems with a large number of modes. In these problems, Gaussian distributions are essential (except at critical points or lines) and the problems are reduced to some excitations of Gaussian modes. In the turbulence, non-Gaussian distributions such as Paretian distributions seem to be essential, as can be typically seen in the Kolmogoroff's spectrum[62] and its intermittency correction.[63-65]

The difficulty of developed turbulence also lies in the lack of a simple and relevant model. A huge computer is necessary to calculate the nonlinear partial differential equations such as the Navier-Stokes equation with a large Reynolds number. Coupled map lattices and cellular automata in Chap. 7 will be of relevance for this purpose.

What aspects are essentially new in high-dimensional systems? In some partial differential equation systems, the states fall into low-dimensional attractors and the systems are described by a low-dimensional chaos theory. In some other systems, the states seem to be approximated by direct product states of low-dimensional chaos. Various important aspects in high-dimensional systems, however, are essentially new. In addition to the aspects in Chap. 7 three general festures are discussed here.

One important aspect is the increase of the number of attractors. In class-2 cellular automata,[66] the number increases exponentially as the system size. For class-4 cellular automata also, it seems to increase exponentially. In the cellular automata with a finite size, any state falls into cycles after some iterations. Thus, we can enumerate all attractors.[67] In the period-doubling of kink-antikinks in coupled map lattices, the number of attractors increase at least exponentially as size.

If the attractor is chaotic without localization, however, it is rather hard to enumerate all attractors. If chaos is localized, a lot of attractors exist. It is not sure how many attractors coexist for coupled map lattices without localization of chaos. In partial differential equation systems, it is much harder to find a large number of attractors numerically.

In low-dimensional systems with chaotic attractors, there is unpredictability in the sense that small uncertainty about the initial condition is expanded. Statistical prediction, however, is possible since there exists only a small number of physical invariant measures.

If the system has a large number of chaotic attractors and the basins of attractions are complicated (e.g., fractal basin structure[68]), even the statistical prediction seems to be difficult. There may exist a huge number of physical invariant measures and small change of initial conditions may cause the state to fall into a different attractor. Thus, small change of initial conditions can induce a different statistical behavior. Class-4 cellular automata seem to have this property.

We do not know how many attractors the Navier-Stokes equation have for turbulence. One possibility is that it has a large number of attractors but that the addition of a small noise can lead to a single invariant measure. The jumping process among attractors by a noise is an important problem to study.[67]

The second problem is about the level of unpredictability. If the disturbance at one position can induce a large change at distant positions after some time (as is seen in §7.7), the system shows a higher level of unpredictability than low-dimensional chaos or localized chaos.

Let us explain this feature by taking an example in weather forecasting. Assume that one wants to know the weather in London after D days with some precision. If the turbulence were localized, then one needs to increase the precision of today's data only around London, in order to increase the prediction. This is the case for localized chaos. Necessary data about London would increase proportionally to $e^{\lambda D}$ (λ: Lyapunov exponent) as D is increased, but we would not need the data about Tokyo. If the turbulence is extended, on the other hand, the necessary data increases not only about today's weather at London but also about at Paris, New York, and even about at Tokyo as D is increased. The area about which the data is necessary increases as D is increased. Thus, the system has a higher level of unpredictability. The relation between the two levels of unpredictability seems to be quite analogous to the relation between PSPACE (the problem that can be solved with polynomial storage capacity but may require exponential time) and NP (both the capacity and time increase exponentially as size) in computer science.[69]

Another problem is nonstationarity and evolution. In open flow problems in §7.5, the pattern changes as the flow goes downstream. There might exist a kind of nonstationarity. As a general problem in a high-dimensional system, nonstationarity will be important. The origin of non-Gaussian distribution in turbulence should also be investigated from a dynamical systems viewpoint. The understanding of the mechanism of the existence of hierarchy of eddies with various sizes may be essential.

When we look back again at the problems in a field theory of chaos, many of them are common to those in artificial intelligence and brain — such as creation, storage, and propagation of information, hierarchy, stability against noise, response, and evolution (see §8.2 (iv)). I would like to close the present section by the next network diagram, which shows my future dream about the nonlinear science.

§8.4 Epilogue

I have put the fascinating figures by M.C. Escher on the title pages of Chaps. 1-7, which might metaphorically suggest important notions in the chapters. On the title page of the present chapter, I have put the famous allegory by the ancient Chinese philosopher, Chuang Tzu. The translation by Burton Watson is as follows (which is given for those who cannot read Chinese letters):[70]

The emperor of the South Sea was called Shu [which means Brief], the emperor of the North Sea was called Hu [which means Sudden], and the emperor of the central region was called Hun-tun [which means Chaos]. Shu and Hu from time to time came together for a meeting in the territory of Hun-tun, and Hun-tun treated them very generously. Shu and Hu discussed how they could repay his kindness. "All men", they said, "have seven openings so they can see, hear, eat, and breathe. But Hun-tun alone doesn't have any. Let's try boring him some!"

Every day they bored another hole, and on the seventh day Hun-tun died.

One could interpret this allegory about the death of chaos in various ways. May chaos be alive in future and keep on revealing marvellous aspects in nature!

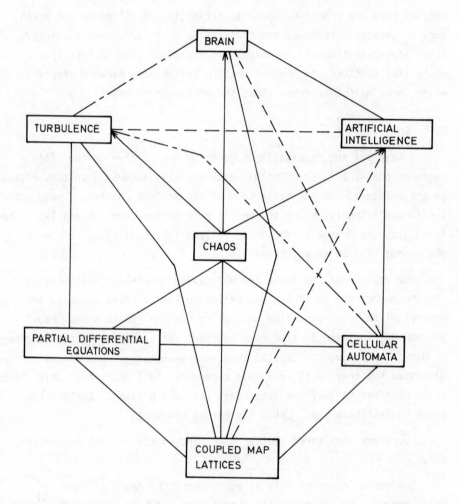

Fig. 8.3 Future dream (or nightmare?) for ABC (A.I., Brain, and Chaos). See text for the meanings of connections.

REFERENCES

1. A self-similarity in a band-merging in a piecewise-linear circle map has recently been found: H. Mori, H. Okamoto, and M. Ogasawara, Prog. Theor. Phys. 71 (1984) 499.

2. For BZ map, see K. Matsumoto and I. Tsuda, J. Stat. Phys. 31 (1983) 87; K. Matsumoto, preprint (1983).

3. J.D. Farmer, Physica 4D (1982) 366.

4. Using the coupled logistic map, we can study the stability of a homogeneous state $(x = y)$ and show that the doubling occurs only a finite number of times for a finite coupling.

5. H. Daido, Prog. Theor. Phys. 71 (1984) 402.

6. F. Argoul and A. Arnéodo, "Scaling for a periodic forcing at the onset of intermittency", preprint (1985).

7. Y. Pomeau and P. Manneville, Comm. Math. Phys. 74 (1980) 189.

8. J.P. Gollub and S.V. Benson, J. Fluid Mech. 100 (1980) 449; J. Maurer and A. Libchaber, J. Physique Lett. 41 (1980) L515.

9. B.B. Mandelbrot, The Fractal Geometry of Nature (Freeman, 1982).

10. J. Kaplan and J. Yorke, in Springer Lecture Notes in Mathematics 730 (1979) 204; H. Mori, Prog. Theor. Phys. 63 (1980) 1044; P. Grassberger and I, Procaccia, Phys. Rev. Lett. 50 (1983) 346; A. Brandstäter et al., Phys. Rev. Lett. 51 (1983) 1442.

11. F. Takens, in Dynamical Systems and Turbulence, ed. D. Rand and L.-S. Young, (Springer, 1981).

12. A.N. Kolmogoroff, Problems in Information Transmission 1 (1965) 3; G.J. Chaitin, Sci. Am. May (1975) 47.

13. The author would like to thank Dr. I. Shimada for useful discussions.

14. See K.G. Wilson, Rev. Mod. Phys. 55 (1983) 583 for a historical review.

15. M.J. Feigenbaum, J. Stat. Phys. 19 (1978) 25; 21 (1979) 669.

16. A critical line appears in Kosterlitz-Thouless transition (J.M. Kosterlitz and D.J. Thouless, J. Phys. C6 (1973) 1181). The critical line, however, has no measure in the parameter space (temperature and a magnetic field).

17. J.D. Farmer, in Fluctuations and Sensitivity in Nonequilibrium

Systems, W. Horsthemke and D. Kondepudi, eds. (Springer, 1984) and Phys. Rev. Lett. $\underline{55}$ (1985) 351.

18. B.J. Alder and T.E. Wainwright, Phys. Rev. $\underline{A1}$ (1970) 18.

19. M. Suzuki and R. Kubo, J. Phys. Soc. Japan $\underline{24}$ (1968) 51; K. Kawasaki, in Phase Transitions and Critical Phenomena, Vol. 2, ed. by C. Domb and M.S. Green (Academic Press, London 1972); P.C. Hohenberg and B.I. Halperin, Rev. Mod. Phys. $\underline{49}$ (1977) 435.

20. See e.g., P. Dutta and P.M. Horn, Rev. Mod. Phys. $\underline{53}$ (1981) 497 and T. Musha, The World of Fluctuations, Kodan-sha (1980) in Japanese.

21. Y. Aizawa, C. Murakami, and T. Kohyama, Prog. Theor. Phys. $\underline{79}$ (1984) 96.

22. B.B. Mandelbrot, J. of Political Economy, $\underline{71}$ (1963) 421.

23. Y. Aizawa, Sugaku Seminar $\underline{263}$ (1983) 36 (in Japanese).

24. R.S. MacKay, J.D. Meiss and I.C. Percival, Physica $\underline{13D}$, (1984) 55; D. Bensimon and L.P. Kadanoff, ibid, 82.

25. D.K. Umberger and J.D. Farmer, Phys. Rev. Lett. $\underline{55}$ (1985) 661.

26. Y. Aizawa, Prog. Theor. Phys. $\underline{71}$ (1984) 1419; T. Kohyama, Prog. Theor. Phys. $\underline{71}$ (1984) 1104.

27. V. I. Arnold, Dokl. Akad. Nank SSSR, $\underline{156}$ (1984) 9; F. Vivaldi, Rev. of Mod. Phys. $\underline{56}$ (1984) 737.

28. K. Kaneko and R. Bagley, Phys. Lett. $\underline{110A}$ (1985) 435.

29. See e.g., M.R. Guevara, L. Glass, and A. Shrier, Science $\underline{214}$ (1981) 1350; H. Hayashi, S. Ishizuka, K. Hirakawa, Phys. Lett. $\underline{98A}$ (1983) 474.

30. In ecological systems, chaos was observed in laboratory, but not in fields: see R. May, Nature $\underline{26}$ (1976) 459.

31. K. Tomita, in Progress of Statistical Mechanics (1981, Syōkabo) (in Japanese). See also Prog. Theor. Phys. $\underline{79}$ (1984) 1.

32. I. Tsuda, Doctor Thesis, (Kyoto Univ., unpublished, 1982) (in Japanese). See also Prog. Theor. Phys. $\underline{79}$ (1984) 241.

33. See, e.g., L. Brillouin, Science and Information Theory, Academic Press Inc., N.Y. (1962).

34. G.S. Stent, The Coming of the Golden Age, Natural History Press, N.Y. (1969).

35. L.B. Meyer, Music, the Arts and Ideas, Univ. of Chicago Press, Chicago (1967).

36. Quite recently I have listened to "logistic map music" by Gottfried Mayer-Kress, which sounds like some electric music, e.g., by Kraftwerk.

37. C.E. Shannon, Bell Syst. Tech. J., 30 (1951) 50.

38. Robert Shaw, Zeit. für Naturforschung 36a (1981) 80.

39. S. Lem, Solaris (Science Fiction: 1961) (Japanese translation by K. Iida, Hayakawa Pub.).

40. N. Chomsky, Reflections on Language, (1975, N.Y., Pantheon Books).

41. J.S. Nicolis and I. Tsuda, to be published in Bull. Math. Biol. (1985).

42. G.A. Miller, Psychological Review 63 (1956) 81; see also, D.E. Rumelhart "Introduction to Human Information Processing", (John Wiley and Sons, Inc., 1977).

43. J.J. Hopfield, Proc. Natl. Acad. Sci. 79 (1982) 2554; 81 (1984) 3088.

44. UNIX: Operating System developed by Bell Laboratory (mainly by K. Thompson, B. Kernighan, D.M. Ritchie).

45. K. Fukushima, Systems Computer Controls 6 (1975) 15; Tech. Monograph (N.H.K.) 30 (1981) 178.

46. D. d'Humières and B.A. Huberman, J. Stat. Phys. 34 (1984) 361.

47. B.A. Humberman and T. Hogg, Phys. Rev. Lett. 52 (1984) 1048.

48. J. Nagumo and S. Sato, Kybernetik 10 (1972) 155.

49. M. Rubinstein and D.R. Nelson, Phys. Rev. B26 (1982) 6254.

50. K. Kaneko and Y. Akutsu, "Phase Transitions in Two-dimensional Cellular Automata", J. Phys. A, in press.

51. S.R. MacKay, A.N. Berber and S. Kirkpatrick, Phys. Rev. Lett. 48 (1982) 767.

52. See e.g., B.A. Huberman, J.P. Crutchfield, and N.H. Packard, Appl. Phys. Lett. 37 (1980) 751; N.F. Pedersen and A. Davidson, Appl. Phys. Lett. 39 (1981) 830.

amura, S. Ohta and K. Kawasaki, J. Phys. C15 (1982) L143;
bson and C. Jeffries, preprint (1983).

ᴋ. Ikeda, H. Daido and O. Akimoto, Phys. Rev. Lett. 45 (1980) 709.

ɔ. A.H. Gibbs, F.A. Hopf, D.L. Kaplan and R.L. Shoemaker, Phys. Rev.
Lett. 46 (1981) 474.

56. See, e.g., M.V. Berry, Annals. of Phys. 131 (1981) 163; in Proc.
on 'Les Houches' Summer School on Chaotic Behavior (1983, North-
Holland); G.M. Zaslavsky, Phys. Rep. 80 (1981) 157.

57. M.C. Gutzwiller, Physica 5D (1981) 183-207.

58. Y. Kuramoto, Physica 106A (1981) 128.

59. R. May, Annals of N.Y. Acad. Sci. 357 (1980) 267.

60. M.C. Mackey and L. Glass, Science 197 (1977) 287.

61. J.C. Eilbeck, P.S. Lomdahl and A.C. Newell, Phys. Lett. 87A (1981)
1; D. Bennet, A.R. Bishop and S.E. Trullinger, Z. Phys. B47 (1982)
265; M. Imada, J. Phys. Soc. Japan 52 (1983) 1946; K. Nozaki, Phys.
Rev. Lett. 49 (1982) 1883.

62. A.N. Kolmogoroff, C.R. Acad. Sci. USSR 30 (1941) 301, 538.

63. E.A. Novikov and P.W. Stewart, Isv. Akad. Nauk USSR, Ser. Geophys.
3 (1964) 408.

64. B.B. Mandelbrot, J. Fluid Mech. 62 (1974) 331.

65. U. Frisch, P.-L. Sulem, and M. Nelkin, J. Fluid Mech. 87 (1978) 719.

66. For the classes of cellular automata, see §7.6 and S. Wolfram,
Physica 10D (1984) 1.

67. K. Kaneko, preprint; see also S. Kauffman, J. Theor. Biology 22
(1969) 437.

68. C. Grebogi, E. Ott, and J. Yorke, Physica 7D (1983) 181; S. Takesue
and K. Kaneko, Prog. Theor. Phys. 71 (1984) 35.

69. S. Wolfram, Phys. Rev. Lett. 54 (1985) 735.

70. The Complete Works of Chuang Tzu (translated by B. Watson; Columbia
Univ. Press., N.Y. and London, 1968).